# Designing MODEL AIRCRAFT

# *Designing*
# MODEL AIRCRAFT

BY PETER MILLER

First published by Traplet Publications Limited in 1995
Reprinted 2002
Traplet House,
Severn Drive,
Upton-upon-Severn,
Worcestershire. WR8 0JL
United Kingdom.

ISBN 0 9510589 6 7

*Front Cover*
*Quark is a CO2 powered model for the CETO sub miniature radio, total, ready to fly weight is 2 ounces.*
*Inset is Pee Tee Sportster for .40 to .45 four stroke engines.*

*Back Cover*
*The author with his unpublished "Belgarion" for Saito 90 twin.*

*Technical Drawings by Lee Wisedale*
*Cartoons by Simon Bates*

TRAPLET
PUBLICATIONS

Printed and bound by Stephens & George Limited,
Merthyr Industrial Estate, Dowlais, Merthyr Tydfil, Mid Glamorgan CF48 3TD

# About the Author

Peter Miller started modelling in 1952 with rubber powered kits before moving on to power models in 1954. After several attempts to learn to fly control line models without achieving one complete lap he designed his own trainer with which he finally taught himself to fly.

Being totally obsessed with aircraft, Peter spent 12 years as as airframe fitter in the R.A.F. which removed some of his illusions about aircraft. As Peter puts it, "Most aircraft are designed by sadists and you need to be a cross between an octopus and a contortionist to work on them."

During this time he continued to build models from kits, plans and his own designs.

In 1974 Peters first plan was published, a control line aerobatic biplane. Over the next 8 years another 9 control line plans found their way into British and American magazines together with several articles.

1982 saw the publication of the author's first R/C design which has been followed by a whole host of varied designs published in most of the magazines in Britain and the USA.

Peter is not only known as a designer, he wrote the "Engine Bay" column in RC Model World for 7½ years and currently has two columns running with other magazines. As well as the columns, Peter has written articles on a wide range of modelling subjects and produced numerous kit and product reviews, as well as writing "Four Stroke Model Engines", the first book on the subject of the modern four stroke engines which was published by Traplet Publications.

# Contents

# Introduction

One of the first questions any spectator asks a modeller is 'Did you make it yourself?' and there can be few modellers who do not feel a touch of pride when they say 'Yes, I did.' Imagine then the pride you would feel it you could say, 'It is my own design.' A response which brings a look of respect to the spectators eye.

Quite apart from the ego factor, designing your own models adds a tremendous amount of interest to the hobby and, not only can it save you money it can actually be profitable.

There are other books on designing models but this book brings the subject down to the simplest possible terms and methods. No maths that cannot be handled on a basic calculator, the minimum of theory which is reduced to the barest facts in understandable terms.

Even if you do not want to design models, knowing a little about design will help you to improve on kits, spot potential weak spots, modify other designs without ruining them and it will help you to understand why things are done in a particular way.

I have not tried to simplify the subject for the book, what I have done is lay out the methods that I use, no more and no less, to produce my own designs.

Judge for yourself how successful my methods are, 48 published plans in 10 magazines in three countries over 20 years at the time of writing with another 8 awaiting publication, and I don't know how many designs that have never been sent in to magazines.

# Chapter 1

# *Why and What With?*

Why bother to design your own models? After all, look at all the kits around. Isn't it enough that your fellow modellers will admire your artistic placing of the stickers on the seventeenth "Maniac" in the pits.

What is wrong with paying £50 for £12 worth of materials in a fancy box, after all, it is all cut out for you and you only have to get the parts to fit. Of course, if you are fussy you might have to replace some of the parts like the one ultra soft fuselage side (or the rock hard one if you prefer).

Want something a little more adventurous? One of the plans published in the magazines, fine, lots of them. I like people who build from plans, it means that the magazines will buy my designs. Actually I wonder why I am writing this. I'm cutting my own throat doing this at the price. Just call me "Cut Me Own Throat" Miller. (With apologies to Terry Pratchett.)

Of course you are limited to building what other people like and there is a good choice but don't you ever want to do something original? Don't you have an urge to create a truly beautiful model. Isn't there a scale subject that you long to build but there are no plans for it? Just once wouldn't you like to have a model that no one else has? If the answer to ALL these questions is "No" all I can say is, "Watch out, the transplant surgeons are looking for spares."

### *Perhaps you think it is hard*

The reason that people don't design their own models is that they believe that it is difficult. Bull . . . Er . . . Codswallop.

I know 16 year old youths who have designed excellent projects for their G.C.S.E., I know of some that have then gone on to sell the plans to magazines. I imagine that most modellers have as much intelligence as a bright 16 year old.

All you need to know to design a model are a few simple facts. If you have built a few kits or the odd plan you already know most of what you need, you just don't realise it.

***The author's first R/C design was this Luton Minor.***

***Tim Rose designed 'Rhubarb' as a school project. Plans then published in RCMW.***

***16 year old Gary Gooderham designed this model for his GCSE project, his first design, flew very well.***

Do you understand the co-efficient of lift or the co-efficient of drag? Do you understand neutral points and polar moments of inertia? Do you study the graphs supplied with the co-ordinates of wing sections? You do? Well, you don't need to read this because I don't and nor will the readers of this book.

You will learn all that you need from this book, true rule of thumb designing and the first rule of thumb is, don't hit it with a hammer because you can't hold a pencil afterwards.

Out of the goodness of my heart and because I am getting paid for it, I am going to tell you how to design models the easy way without any of the mystery or formulae.

## *What you need*

You can design a model with a pencil, a ruler, a table and a sheet of brown paper and an eraser (don't forget the eraser). It isn't very easy but it can be done.

At the other end of the scale you can have an AO drawing board with parallel motion, a full set of drawing instruments, French curves, assorted templates, several pencils and pens and Mylar film at £50 a roll. All very nice but it won't make your designs any better, easier perhaps but not better.

Most people will start out with a bit more than the most basic and slowly build up as they progress. Let's look at a simple and cheap kit of equipment.

You need a drawing board, it must be big enough to draw a complete fuselage of the size you prefer and at least one wing panel. You don't have to do everything on one sheet so it doesn't need to be that big. Let's say about 48" by 36". You might even be able to use the back of your building board. You can buy a nice piece of ½" plywood or blockboard for about £4. Make sure that the edges are square, sandpaper it smooth and you have an excellent board, I used one like that for many, many years.

You need a Tee Square, a cheap one will be fine, if you can get one as long as the board, good, but a shorter one is OK.

You need a couple of rulers, a Metre rule which must be straight. You can find them in Woolworths or DIY stores at very sensible prices (under a fiver) but check for straightness or you will have funny shaped wings and strange incidence angles. You also need a 12" or 24" plastic ruler, both is even better.

A good large set square is vital, get a 12" one as this will enable you to extend lines from a side view right up to a plan view as we shall see in a future chapter. I also use a square 5" protractor, this is a wonderful tool and can be obtained from pilot's suppliers as it is used for navigation. It will cost about £4, it isn't essential but it is extremely useful and you do need a protractor.

Pencils are as essential as a modelling knife is to building. Use a 2H as this will give a dark enough line but will hold its point, buy a good sharpener to go with it. You can buy Pentel propelling pencils with different size leads, 0.50mm is about right, they are not expensive.

You will need a good pencil eraser, get a nice big soft one, a quality one, the nasty little things sold for school use will soon get dirty and leave dirty smears on your drawing, an untidy drawing is not as nice to work on and can lead to mistakes.

A French curve is useful, you can buy sets but I find that I only use one (see fig 1.1), this is quite expensive but cheaper ones which are supposedly the same shape do not have quite the same flow from curve to curve. You will need this for airfoil sections.

A pair of compasses is needed for drawing wheels and the front of spinner rings. One that can be converted to dividers is useful. Get a reasonable pair, not one of those things that you stick a complete pencil in.

***First model design by retired full size designer, over strong but not too heavy, needed more power but was nice flier.***

You now have a set of equipment that will produce almost any design that you want. There are one or two other items but they will be mentioned later and either don't cost much or alternative methods can be used.

You should be able to get all the above equipment for far less than a cheap kit and you will save that with your first design.

It is worth going to car boot sales and auctions where you can sometimes find items. My AO drawing board complete with parallel motion and stand came from a colleague via a car boot sale, it cost me 3 ounces of tobacco.

***Your ruler must be straight***

*Fig 1.1 – The most useful French curve*

***My favourite shape of French curve. Figure is to illustrate what you can do with French curves . . . maybe I should re-phrase that!***

## Paper

Drawing paper and tracing paper can be expensive when bought on a roll or in sheets, but you are going to be drawing, correcting and altering, and quite often you will tear a fuselage side view off the board and start again, then, when you have done the drawing you are going to be building on your plan so you want something cheap and tough.

One source of paper is the back of other plans, the ones that you bought because you thought that you might build them only to find that they were designed by a sado-masochist who would not use one piece of wood when he could use ten.

For a really good source of drawing paper go to a DIY store and find their lining paper, tough off white paper and the price of a roll will be cheap . . . I pick mine up from a local shop at about 40p a roll.

Another type of paper which is harder to find is white shelf lining paper, this is thinner and smoother, I like it but can rarely find it these days.

Once you are into designing you can go on to "Detail" paper, this is a thin white paper which will also act like an opaque tracing paper, it does cost nearly £20 a roll though.

In future you may also want tracing paper so that you can have prints run out from your plan, it also allows you to keep a copy while you build on the scruffy original, this is useful it you want to sell the plan to a magazine after the test flights.

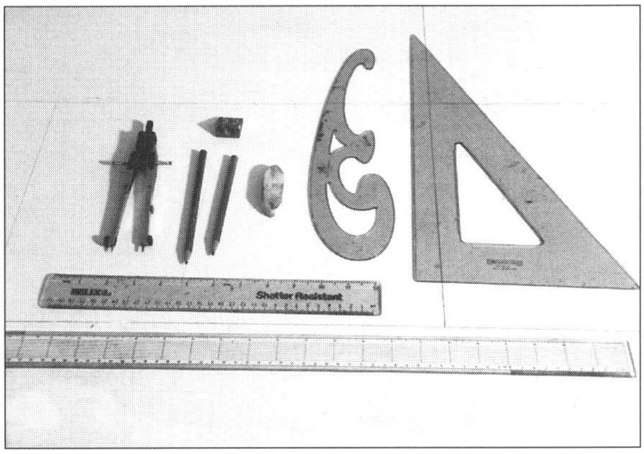


***The basic needs in the way of drawing instruments.***

***The writer's bargain drawing board, sheer luxury after years with pieces of plywood.***

# Chapter 2

# Essential Aerodynamics

Aerodynamics is a vast and complicated subject but the basic theory of flight is much simpler, and what you need to know to design a model is even simpler than that.

I knew a University graduate who designed a series of beautiful control line stunt models, all the formulae, the works and they flew very well indeed. One day he needed a stunter in a hurry and designed a simple box and plank design for quick building . . . it flew far better than the fancy ones.

Don't think that all the computers and wind tunnels and expertise of the full size world are infallible, they get it horribly wrong as well at times.

I am going to tell you the minimum that you need to know to design a model, if you want to understand aerodynamics better I can recommend "The Mechanics of Flight" or "Flight without Formulae" by Kermode, old books but still two of the easiest to understand. They should be available from your library or aviation book shops.

## All you want to know about flight

An aeroplane flies because the wing is moved through the air and generates lift, the lift is generated from one or both of two actions.

Reason one is that because the shape of the cross-section (called the airfoil from now on) makes the air travelling over the top surface move faster than the air moving under the lower surface, this reduces the pressure of the air above the wing.

The second reason is that the wing meets the air at an angle forcing air down and therefore the wing is pushed up. This angle is called the angle of attack.

I could go into details as to why this is and explain it all but you don't need to know that and it will take up valuable space.

The wing needs to be stabilised in flight and for this we use a tailplane, we could use a foreplane and build a canard or even a flying wing but that would be running before we can walk.

***Tipsy Nipper is highly aerobatic, note unusual wing section and large incidence angle compared with tailplane and engine thrustline.***

We also need a power source to propel the wing through the air. On a glider this is simply the weight of the aircraft, it is literally sliding downhill all the time. Yes I know gliders can gain height, look at it this way, if you are in a lift and launch a paper glider which takes five seconds to reach the floor from six feet and the lift is going up at ten feet per second the glider will have gone up 44 feet in the five seconds it took to reach the floor.

There are four forces acting on an aircraft; weight, lift, drag and thrust. When the aircraft is in stable flight, i.e. travelling at a constant speed in a constant direction, these forces will balance out. Trust me, they do and you don't really need to worry about that at the moment and I will explain as and when you do need to know.

## *The Airfoil Section*

I have said that the shape of the airfoil produces lift and it does but there are hundreds of airfoil sections, which do we use?

For our very simple minds we can divide them into four types. Flat bottomed, semi-symmetrical (or Bi-convex as the purists call them), symmetrical and under cambered. Flat plate can be considered symmetrical and curved plate is under cambered.

These are rather inefficient sections for wings and will only be mentioned in passing, although flat plate sections are very common for tailplanes and the flat plate is also used for very simple models with a solid sheet wing.

***Fig 2.1 – Airfoil Types***

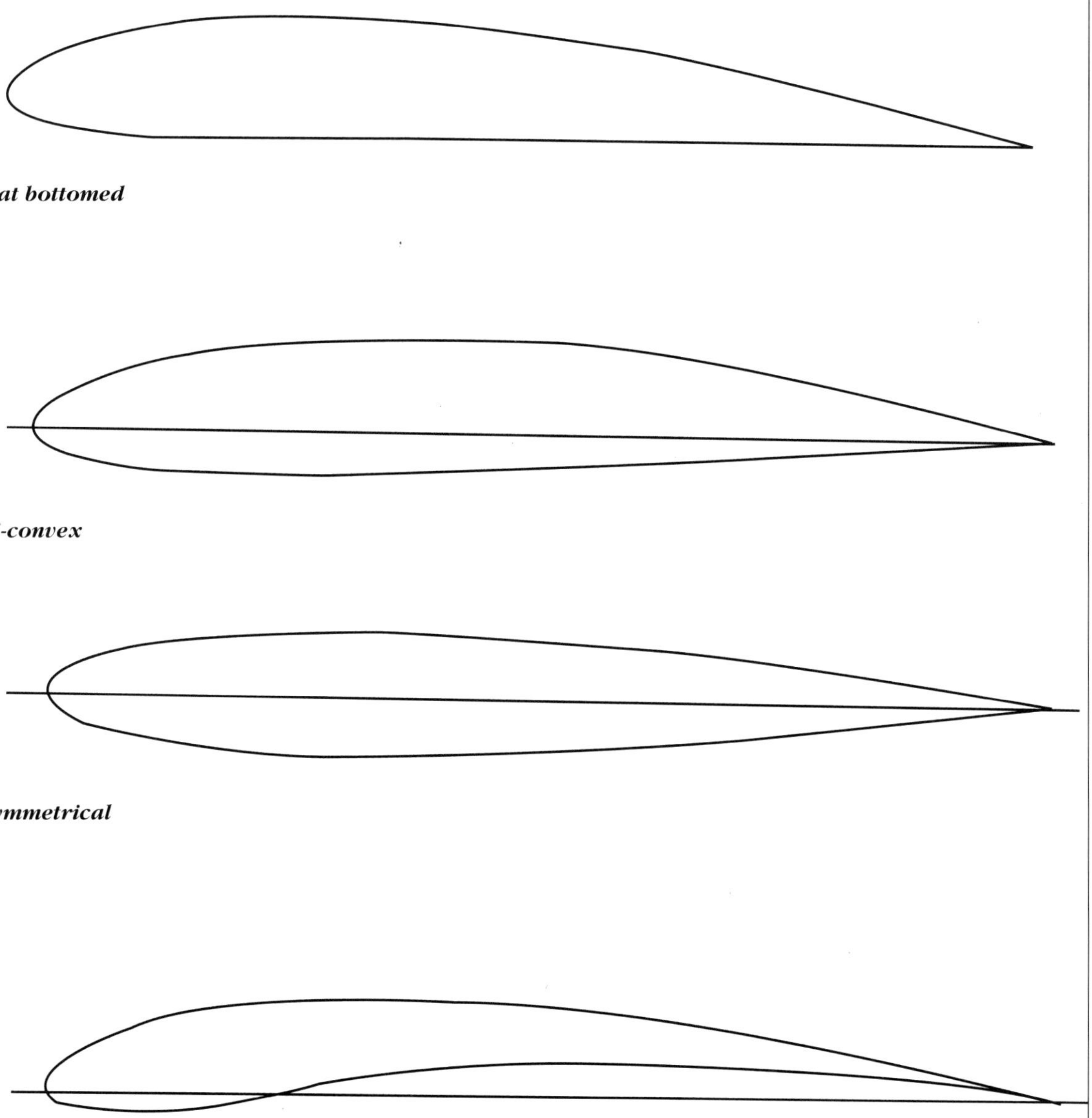

***Flat bottomed***

***Bi-convex***

***Symmetrical***

***Under cambered***

***High wing cabin models can be sleek and beautiful, the author's Lady Eowyn design which is in RCMW plans range.***

Flat bottomed sections are used for some trainers and are quite common on certain scale types, Clark Y is the great classic flat bottomed section. They are easy to build because they can be placed flat on the building board.

Semi-symmetrical sections are very common, they are used on many full sized and model aircraft and are good for the faster trainers and sports aerobatic models.

Full symmetrical sections are for true aerobatic models as they will fly in exactly the same way upside down as the right way up if they are set at the correct angle.

Under cambered sections are used on vintage models and some floater type gliders, they are for slow flying models only as they do produce a lot of drag but they also provide a lot of lift.

For our purposes we will only use the first three types of section.

## *Angles of D . . . Different Sorts*

In describing how a wing produces lift (and incidentally, giving all aerodynamics experts apoplexy) I mentioned 'Angle of Attack', forget about it. The angle of attack is constantly variable and the only time you need to worry about it is when you exceed the angle when the wing stalls and even that varies with speed, but we are not talking about flying so, as I said, forget it.

The angle that we are concerned with is the angle of incidence. This is the angle of the wing to a purely arbitrary line called the datum. The datum is normally taken to be a centre line drawn through the fuselage to which all other angles are related. You could have a datum line at 45° to the fuselage but it would make drawing difficult.

***Fig 2.2 – Terms Explained***

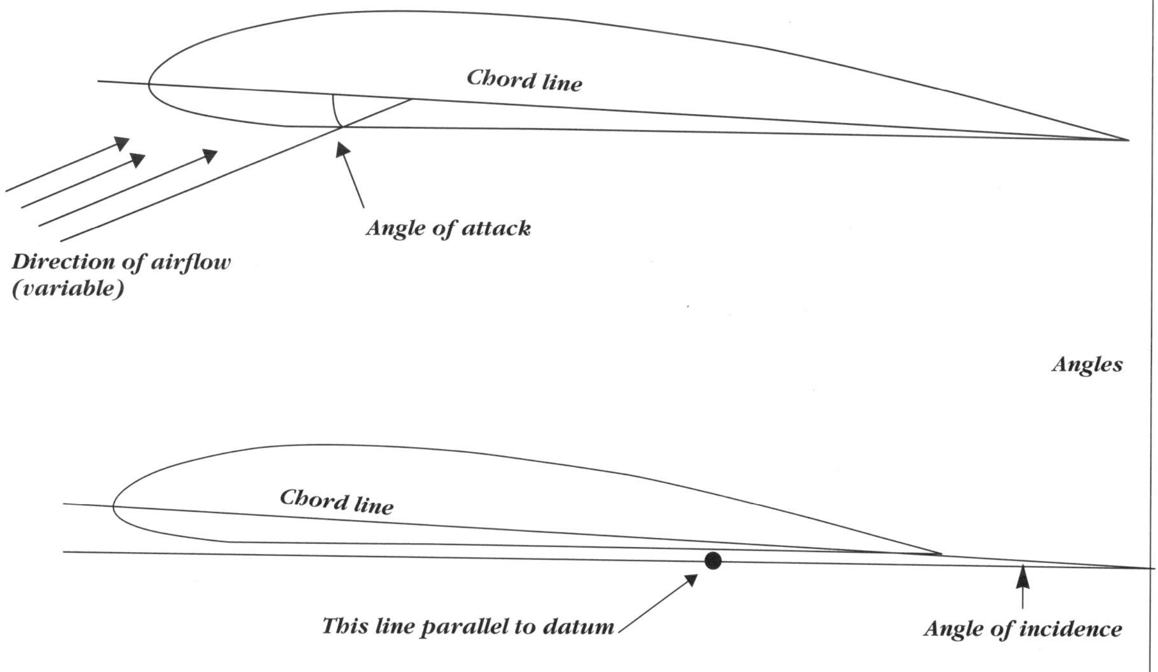

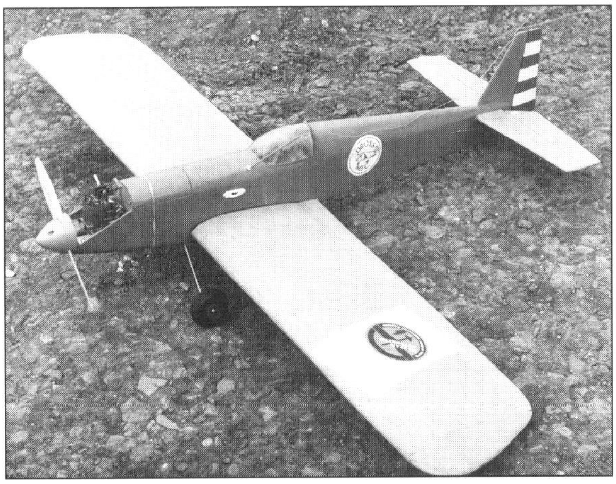

***This model is based loosely on control line stunter layout, flew well. Four strokes are best in light models.***

The angle of the wing is measured from the chord line and the chord line is a line drawn from the centre of curvature of the leading edge to the centre of the trailing edge. Sometimes the chord line is taken as running along the bottom of a flat bottomed wing.

Other items which are set at some angle in relationship to the datum line are the thrust line and the tailplane chord line.

The only other angle that we are concerned with is the dihedral angle which affects stability and aileron response, but this will be discussed when we get down to putting pencil to paper.

## Wing Areas

The wing area of models varies depending on the type of model. Powered gliders will have large wing areas for very small engines.

My Falcon design has 620 sq. in. and is mildly aerobatic on a .12, while Tequila Sunrise has only got around 230 sq. in. for similar power and is fully aerobatic but naturally does not have the same low speed performance or glide.

The model with the smaller wing will have better penetration in a wind because the drag is less in proportion to the power. Fit a .40 to a Falcon and it would handle even stronger winds, but it would then have to be structurally stronger to take the power and suddenly it is not a powered glider any more.

I have given a table of suggested wing areas for most sizes of engine. This table is for average sports

***Full size Skybolt and model Volksplane obey the same rules for flight.***

aerobatics models which is the most popular type of model and is only intended as a guide.

Biplanes will have more area, some light types of scale models will have much more area and powered gliders even more.

Going below the smaller areas would not be a good idea for any type of model as it would be very difficult to keep the wing loading within the bounds of reason.

| | |
|---|---|
| .09 – .15 | 200 to 350 sq. in. |
| .20 – .25 | 350 to 450 |
| .29 – .40 | 450 to 550 |
| .40 – .45 | 550 to 650 |
| .45 – .60 | 650 to 800 |

***Small, ultra simple design can be built quickly and still provides fun and satisfaction.***

## *Areas and Loadings*

One of the most important factors in any design is the wing area and the wing loading. Too high a wing loading and you will have a model that is a pig to fly, if it will even get of the ground.

This is where we have to do a little maths, nothing complicated, you don't really need a calculator unless you are a product of the modern education system.

Wing area is quite simply the chord or width of the wing multiplied by the span or length of the wing. Egad! How basic can I get!

Wing loading is the figure that you get when you divide the estimated weight of the model in ounces by the wing area in square feet.

Let's take an example. The wingspan is 50", the chord is 9" so: 50 x 9 = 450 square inches. Divide by 144 which is the number of square inches in a square foot and that is 3.125.

Assume that we are designing a model for sports aerobatics with a wing of the size given above, if we build fairly light it should come out at about 4 lbs in weight. 4 lbs is 16 ounces x 4 = 64 ounces.

If we divide 64 ounces by 3.125 square feet we get 20 ounces per square foot wing loading.

After you have built the model and found out what it does weigh you can see how close you got.

The same system will allow you to decide on how big a wing you need. Assume that you are designing a model for a .40 engine, a realistic weight for this would be up to 6 lbs maximum. A sensible wing loading would be up to 24 ounces per square foot.

6 lbs = 96 ounces. 96 divided by 24 equals the desired number of square feet = 4. Multiply by 144 = 576 square inches. That is the size of the wing but we can bring it up to 600 for safety.

I had planned to include a table of suggested wing

***Competition Fun Fly models take low aspect ratio and low wing loading to extremes for incredible performance.***

loadings for different sizes of model, but when I began to lay it out I realised just what a wide range there are, even for one size of engine.

For example, a Competition Fun Fly model might be powered with a .25 to .40 and have 700 square inches of wing area, a wing loading of around 10 ounces per square foot, but a sports model might go up to 25 ounces per square foot which would definitely be the top limit for the .25 size of engine.

Lets look at some scale models. The Sig Piper Cub is 72" span and can be built to weigh around 4lbs, it flies like a dream on a 20 four stroke. The Great Planes Piper Cub is the same size, but they claim that their kit will weigh 6½ to 7½ lbs and needs a .40 to .60 two stroke.

The only comment that I will make is that the real Cub is light and flies very slowly.

Running through some of the figures for my own designs, I suggest that, for normal sports aerobatic types you do not exceed 20 ounces per square foot for models up to .15 engines and 22 ounces per square foot for .25 size models, up to .40 you can go up to 26 ounces and for .60 you might go as high as 30 ounces per square foot.

Going a little over these figures will not be disastrous, but keeping below them will help improve performance and handling of the model.

***Sailplanes use high aspect ratios, this Condor MS has an A.R. of about 20.***

Powered gliders, lightweight scale models like Piper Cubs etc. should have much lower wing loadings.

***Fig 2.3 – Aspect Ratio***

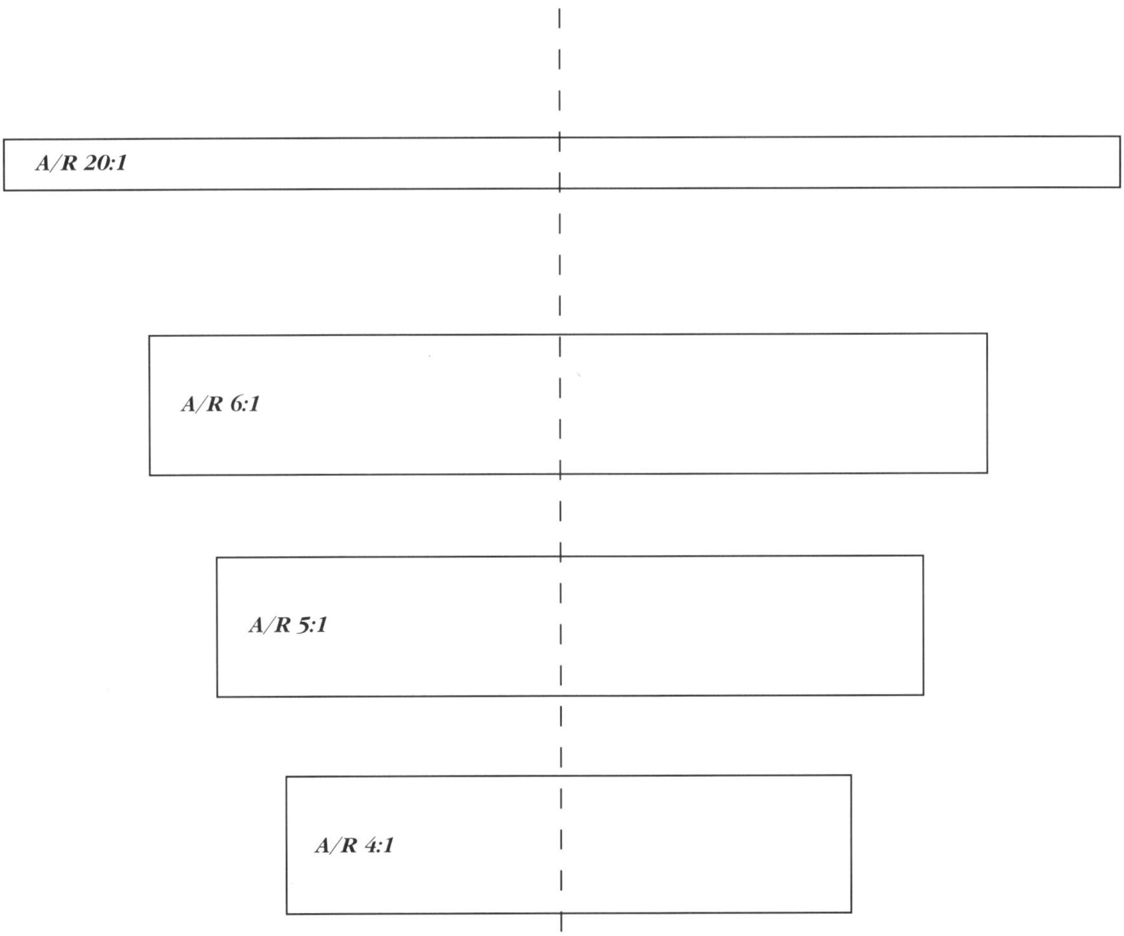

***Full size Kitfox is microlight, very like a model in construction.***

## *Wing Volume Loading*

Giant scale models use much higher wing loadings and if you start working out the loadings for full size aircraft they become astronomical in comparison with their model counter parts. This is because of the claimed extra efficiency with larger sizes.

Many years ago Dave Platt did some comparisons and found that if one took the volume of a wing as a unit of loading instead of the area and compared that at different scales of a particular model of a subject the figures very closely matched.

Dave's figures showed that increasing the size of a scale wing from 500 square inches to 700 sq. in. increased the area by 40%, but increased volume by 98%.

Dave went on to show that a loading of 1 ounce per 5 cubic inches was about right for a heavy scale model and that this coincided nicely with the heavier loadings possible with larger models.

It would seem likely that if this was taken to it's logical conclusion, the volume loading on a model would be very close to that of the full size prototype.

I don't suggest you try working the volume out, but it does explain why larger models can use higher loadings.

Presumably the computer "Whizz Kids" could write a programme that would calculate the desired figures but for our simple purposes it would be very complicated bearing in mind that each aerofoil section would give a different volume.

## *Aspect Ratio*

Yet another of the secret terms of the designer. All it means is the shape of the wing, for a parallel chord wing it is simply the wing span divided by the chord. Example, from the above, 50" span divided by 9" chord = 5.55, a nice figure.

Normal range of aspect ratios for parallel chord wings is about 4.5 to 6. Gliders use A.R.s as high as 20:1.

For a tapered wing we add the tip chord and root chord, divide by 2 which gives the average chord and then divide that into the span.

## *Balance Point or Centre of Gravity*

One of the most vital things about a model is where it balances. This balance point or centre of gravity (CG) will affect the stability to such an extent that if it is just a little too far back it will be impossible to fly the model.

The CG has to be in front of the point through which the lift acts. Now working out this latter point is not only complicated but rather hit or miss as it depends on so many factors so we adopt a rule of thumb for the time being.

On conventional models the CG should be between ¼ and ⅓ of the chord back from the leading edge of the wing, 25% to 30% if you prefer and yes, I know that ⅓rd is more than 30%. The nearer the forward limit the safer the model.

# Chapter 3

# The First Design Layout

In Chapter two we looked at the basic requirements of an aircraft, we talked about wings and angles and balance points and so on, the aerodynamic side of things.

The type of model that we want will decide the aerodynamic layout and the aerodynamic layout will have some considerable effect on the shape of the model, thus all basic trainers are high wing and all pattern ships are mid or low wing. Now you know why every airliner looks the same as all the rest and why modern owners of full size sailplanes can't tell one from the other.

The reason that military aircraft all look different is because they are drawn to specifications decided by committees who don't know what they want anyway.

What shall we design? Well, I imagine that you are past the trainer stage and the most popular type of model is the sports aerobatic type so that is the way we will go. It can be for .25 size engines or .40s, I will give figures for both but I suggest the smaller size as it is cheaper to build.

***Ultra boxy camera plane is the easiest type of model to design. This model would use flat bottomed wing or semi-symmetrical section.***

***Fig 3.1 – The Three Most Useful Sections.***

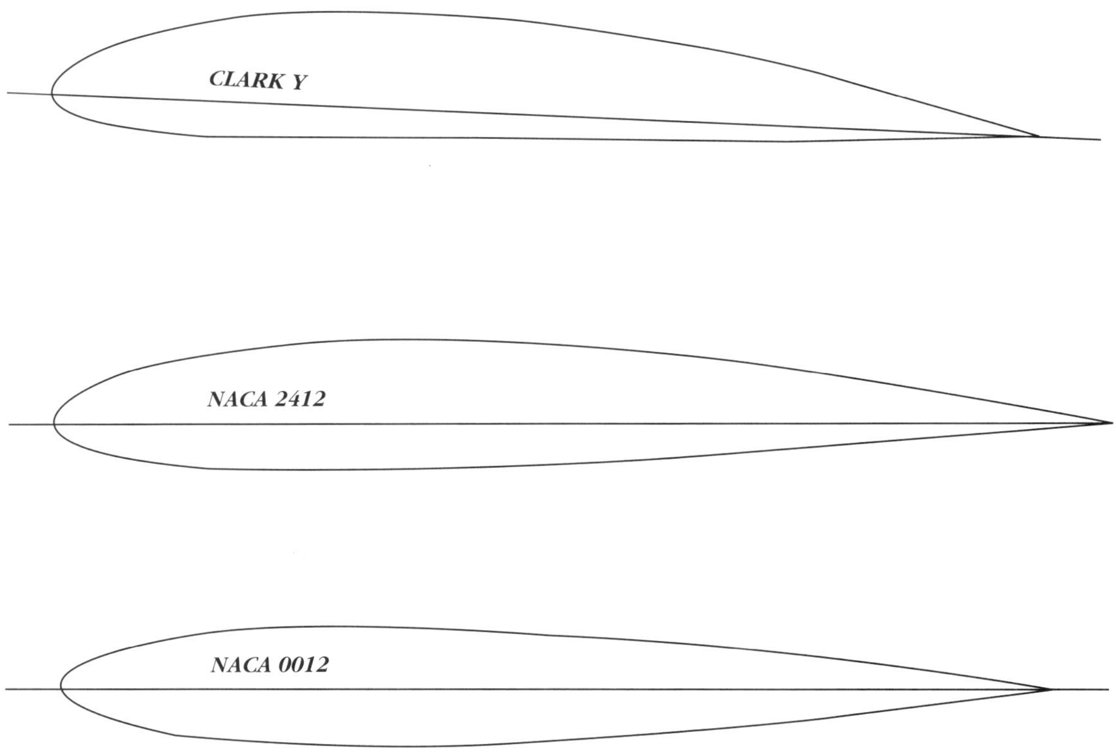

## *Specification*

*Experimental design by George Strohm does not have a single curve in it, every wing section is a diamond. Flew at high speed but was diabolical at low speed.*

Before we can start drawing we need to decide on the size of the model, the wing section that we want to use and the shape and position of the wing.

For a .25 size model a wing of about 450 sq.in. is about right. Now we play about with span and chord until we get an aspect ratio that is sensible, I took 50" span and 9" chord for an aspect ratio of 5.5 but you could vary this from a span of 45" and chord of 10" (A/R 4.5) to a span of to 52" and 8.6" chord (A/R 6). The shorter wing span will roll faster.

You can round figures up and down, a little more area or less is nothing to worry about. For a .40 size model the area would be 550 or more sq. in.

The wing can be located in the shoulder, mid or low positions. The mid-wing presents structural problems and takes up valuable space, the shoulder wing is nice but can make for extra work later, it is ideal when the model has to be hand launched and also makes working on the model easy as it will sit on its undercarriage.

The low wing model is more common and has many advantages but may present problems when it comes to locating the undercarriage in the ideal position. We will cover both shoulder wing and low wing in the exercise, after all, this is your first design and you are going to build it. We also need to decide on the wing section. Here the choice is bewildering. You could use a semi-symmetrical wing or fully symmetrical, you could even use Clark Y set at true ° incidence as this works like a thin semi-symmetrical wing inverted but it is a bit thin.

I favour a thick symmetrical section with a blunt leading edge, this will slow the model slightly and gives very gentle stall characteristics.

In this chapter you will find three airfoil sections, you can have these enlarged on a photocopier to the size that you need, take the measurement of your wing chord and divide that by the measurement of the section printed here, that is the enlargement factor to set on the copier which must have a full zoom capability. e.g. Your model chord 9.5", in the book, 5.375, you need to enlarge by 1.76.

You could design you own section, I have shown the main points that you need to watch. If you go this route you will probably not notice the difference. You could even use the MILLER T.L.A.R. (that looks about right) section.

## *Paper and Pencil*

Take your piece of paper and attach it to your drawing board, I suggest masking tape at the corners but drawing pins or even proper drawing board clips can be used.

Now draw a line about 10" inches from the bottom of

*Fig 3.2 – Own Design Section*

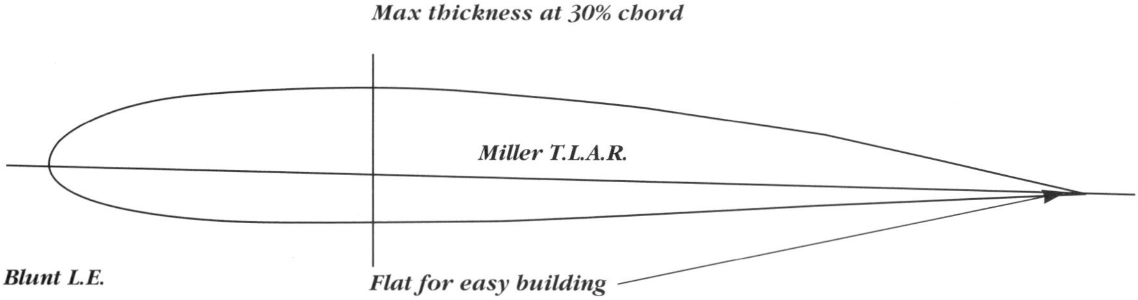

the paper right along. This is your datum line and the whole model will be drawn round this.

The datum line is arbitrary, it can be on the thrust line or it can be the top of the fuselage side, it could even be outside the model but this is not wise. We will make it the top of the fuselage side although you won't know it at this stage.

About 12" in from the left side of the paper drop a line down from the datum, this will be the L.E. of the wing.

Now we draw in the chord line. If this was a trainer we would put it well above the datum line, for a shoulder wing model we can leave it on the datum and or a low wing model it must, naturally, be below. Put it about 3½" below and parallel with the datum.

Now, on the datum, measure back from the L.E. the distance that you have decided on for the chord and drop another line, this is the T.E. Note that you always work on the datum and drop or raise lines.

We now have two more decisions to make, the length of the nose and the tail moment arm or the distance from the T.E. of the wing to the L.E. of the tailplane.

The length of the nose depends on the weight of the engine, the weight of the rear end and the length of the tail moment and you can't work it out so, for most types of model, we can work to a distance from the prop driver to the L.E. of about one chord. Because four strokes are heavier you could reduce it a bit if you are going to use one.

That chord dimension is a very handy one for giving proportions for any size of model.

The tail moment arm and the size of the tailplane are linked with stability. A long moment arm and small tailplane give the same stability as a short arm and big tailplane, however, control is smoother with the longer arm.

A good compromise moment arm is about one and a half chords with a tailplane about 18% to 20% the area of the wing. Pattern ships have longer arms, some of the wilder fun fly models use less.

Having decided on these dimensions we drop a line down 9" in front of the L.E., this is the front of the model and we drop one down about 13" behind the T.E. this is the L.E. of the tailplane.

The tailplane sizes have to be calculated now. Percentage of the wing area is easy, divide the wing area by 100 and multiply by what ever percentage you want. E.g., we have a wing area of 450 sq.in. so 450 divided by 100 equals 4.5. If we go for 18% we multiply by 18 and get a tailplane area of 81 and so on, most calculators have a percentage button which makes life even easier. I prefer a little more area so we will try 20% and get 90 sq.in.

We now want to know the chord of the tailplane, not quite as simple, we may have a parallel chord wing but the tailplane is much nicer if it is tapered. Play with your calculator again. Hit 90 and divide by 5, that gives us a span of 18" and an average chord of 5". Doodle around on some scrap paper and see what looks nice, a taper of two inches looks good, right, root chord of 6" and a tip chord of 4". You can vary this and you can put more taper on the L.E. or on the T.E., just keep the area about the same.

***Fig 3.3 – Basic Layouts***

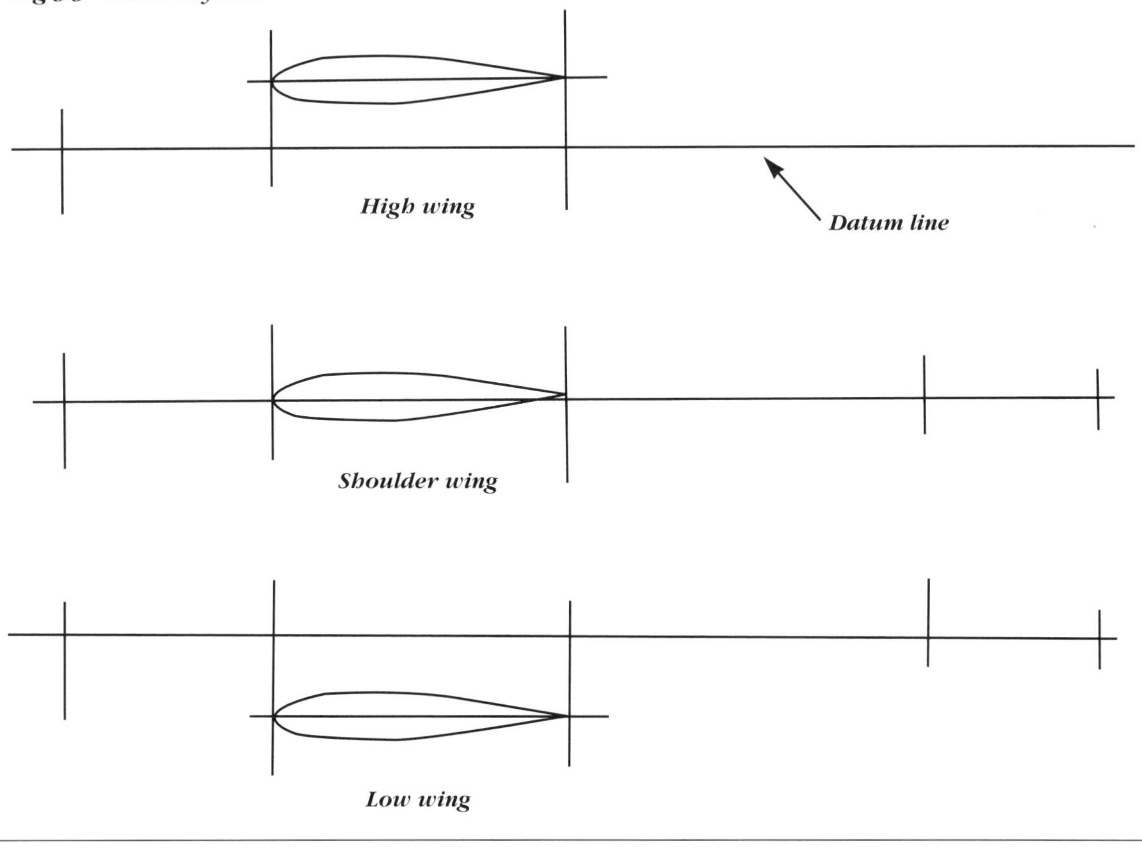

*Fig 3.4 – Aerodynamic Layout*

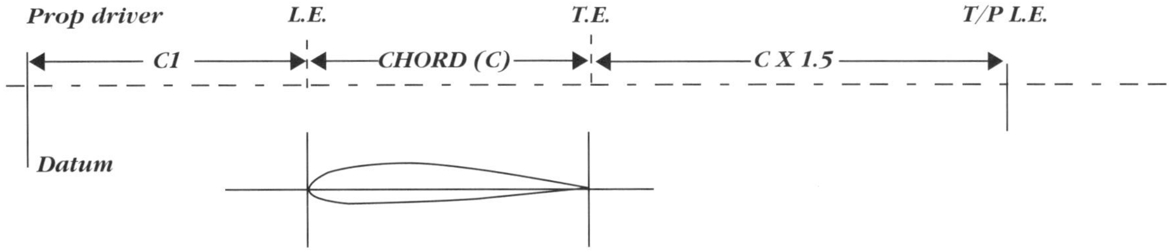


Put the tailplane's 6" root chord on your datum line. You can decide where the tailplane will sit now, I like putting it on top of the fuselage side.

## *Variations on the Theme*

What you have drawn above is the aerodynamic layout for a sports aerobatic model. With a model of this type the wings, tailplane and engine will normally be set at 0° to the datum, this will give neutral stability or, in other words, the model will stay pointing in any direction that you have placed it in. Some other types need to be different.

Trainers need some positive incidence on the wing to make them pull out of a dive if the learner releases the stick. There are so called trainers that have neutral stability and some that are aerobatic . . . they are not basic trainers whatever the box top says.

We talked briefly about the balance of forces and drag in Part II, now we need to look a little further.

Take a high wing model, the wing produces lift but also drag and it is high. The engine is lower down and produces thrust, as you can see from the drawing this tends to pull the nose up so we tilt the engine down and this counteracts the imbalance of forces.

Likewise, it we mount an engine on a pylon, on a seaplane or a powered glider for example, the thrust is well above the drag and so the model will tend to nose dive and in this case we would use up thrust.

Quite honestly, it is impossible for the average modeller to calculate the exact angles needed and so provision must be made for adjustment during the flight test period.

***High mounted pusher engine on the Polish homebuilt needs some 'up' thrust but, because it is pushing the engine is pointed downwards slightly.***

*Fig 3.5 – Thrust Line Adjustment*

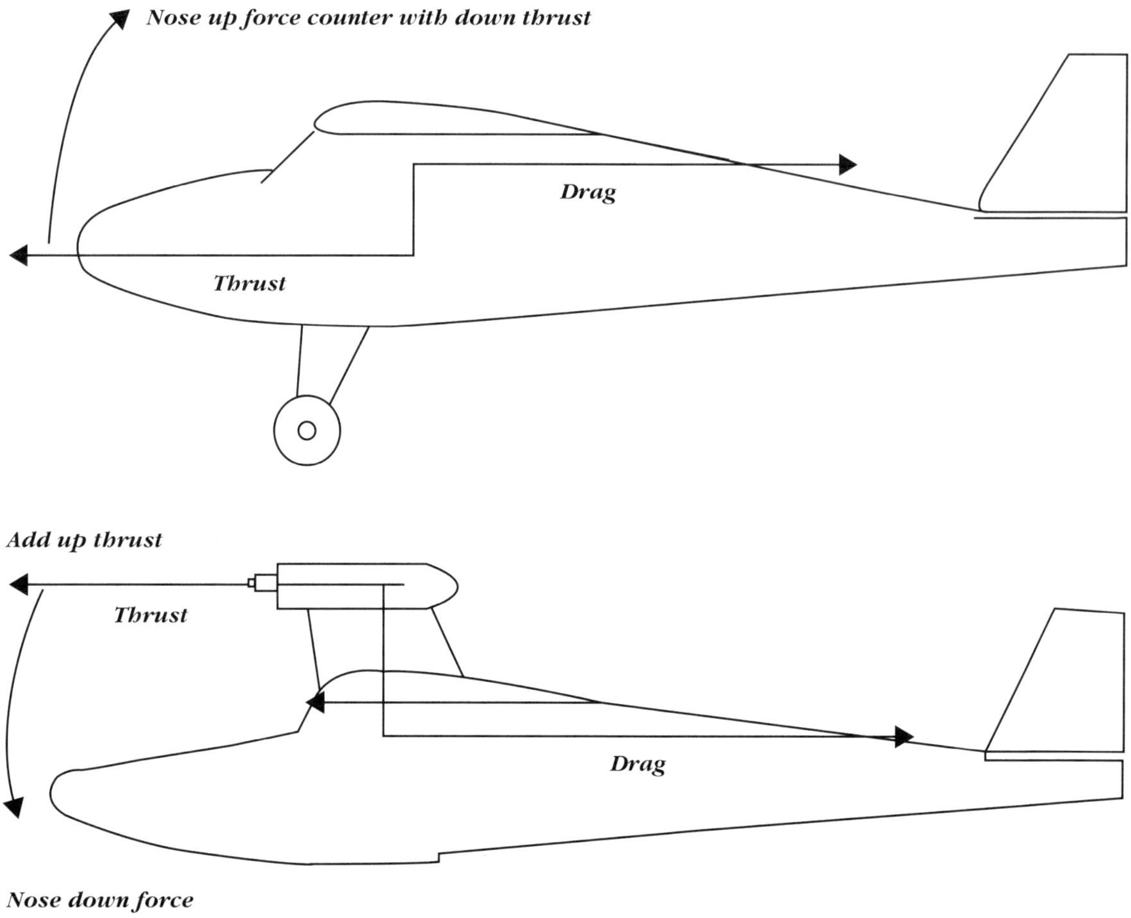


## *Fitting the Gear*

It may sound strange to talk of fitting radios and tanks when you don't have more than a few lines on the paper but now is the time to plan it, before you are restricted by shapes.

Start off by placing your wing section on the plan, if you have copied a section from the chapter you can stick it down on the drawing, just extend the chord line and the line dropped for the L.E. and use these to line the section up with your drawing.

Now you need to draw in the amount that your aileron servo will stick up above the wing. I find that cutting out card shapes the same size as my servos, receiver and battery pack is useful, you can juggle them around and be sure that all the control runs will clear each other.

Don't forget that the aileron servo will be behind the main spar which is normally about 30% of the chord back from the L.E.

You also have to locate the engine at this point. Is it going to be upright or side mounted? You can also position the fuel tank, this should have its centre level with or just below the centre line of the carb. You must allow room for the engine mount and a former between the tank and the engine.

You do not have to draw these parts in detail, just an outline is fine.

Once you have worked out where all the internals will go you can, at last, draw the side view of your model. Here you can let your imagination have free rein, racer, jet style, whatever you like.

I would suggest that you keep it simple at this stage, you still have to draw all the other parts and then build it.

## *Start with a Box*

90% of all model aircraft are based on a plain box fuselage. Some, like the Stik family don't go any further, others have a simple turtle deck and others have the box hidden under stringers and fairings.

Your box will look something like the drawing, the slope up to the tail may vary but at this stage in the learning curve little else will. What you do with it is up to you, a simple turtle deck will be easy to design and can be varied in height, position of cockpit and so on.

Draw it in lightly, this is when the eraser comes into its own as you change things until you get a shape that you like. Of course you should have had a fair idea of what you wanted from those sketches on the back of envelopes, but getting it to look right at full size can take a little time.

A very simple jet style model is shown in Fig 7 this uses the box and a few pieces of sheet to create an attractive looking model.

***Fig 3.6 – Equipment Layout and Outline***

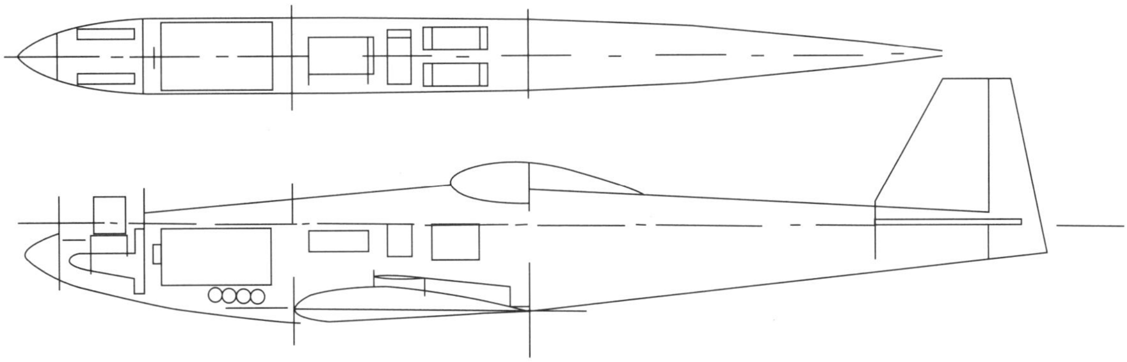

You can draw in the fin, this should have about half the area of the tailplane, more if you have a very bulky front end of the fuselage. Fin shape is not critical, you can have any shape you like, just keep the areas correct and remember that part of it has to be hinged for a rudder.

The rudder area is not critical, highly aerobatic models will benefit from large rudders to kick them over in good stall turns and produce crisp flick manoeuvres and spins, sports model do not need as much. A good compromise is about ⅓rd the total area of the fin and rudder. The rudder is normally extended down to the bottom of the fuselage.

With the side view drawn we can draw the plan view. Draw a line above or below your side view, this line must be parallel with the datum line. Now drop lines from the various critical locations like the front former, the L.E. and T.E. of the wing and the tailplane L.E. and the end of the sides.

You now have to decide on the width of the fuselage, this will depend on the size of the servos and tank. Always leave some room to play with and don't forget to allow for the thickness of the sides and any doublers.

The fuselage can be parallel from the nose to the wing T.E. and tapered from there to the end. You can

***Planning installation during designing will save a lot of problems later.***

*Fig 3.7 – Styles on same Basic Box*

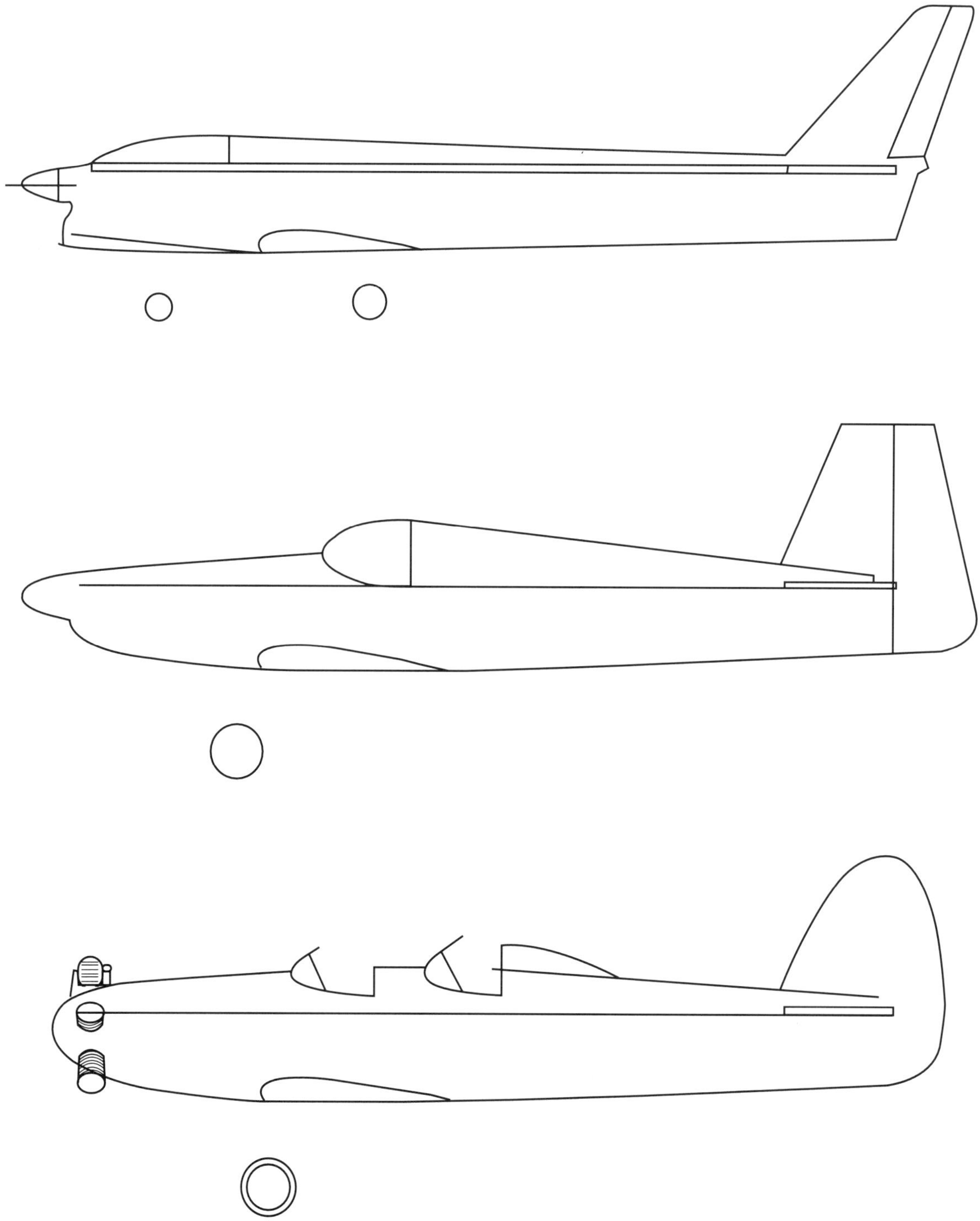

use a smooth, natural curve or a tighter curve and straight taper.

To draw a smooth sweeping curve you need a spline, this is not expensive because it is just a strip of ⅛" square balsa or spruce. Using masking tape hold one end down to the straight portion of the fuselage and then pull the free end to line up with the centre line at the rear of the fuselage.

A light pencil line along each side of the spline and you have the plan view of a curved fuselage side drawn.

The alternative is to use the French curve to draw a gentle bend behind the wing and then add the straight taper back to the end.

The only other task at this stage is to measure down from the thrust line a distance, that will allow plenty of clearance for the intended propeller and draw a line parallel with the datum, this is the ground line and will define how long the undercarriage is.

That takes care of the fuselage outline.

## The Wing

The wing can be drawn as a rectangle because we are not going to go in for tapered wings at this stage. Straight wings are easier because all the wing ribs are the same with just minor variations.

*Make sure the radio will fit*

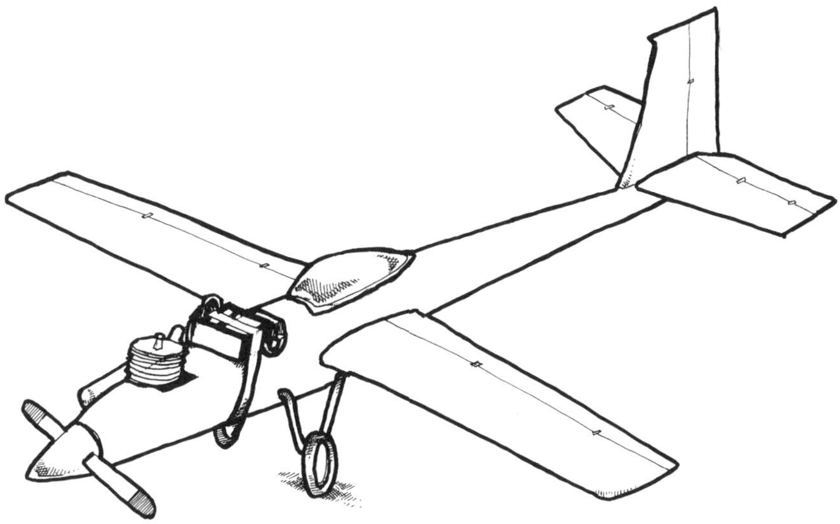

At this stage we can design the shape of the tips and we can also plot the ailerons.

The tips can be any shape that you want but you really want to keep them simple and light at this stage. The shape shown on the drawing is as good as any for a first model.

For ease of construction strip ailerons are hard to beat. I have found that 14% of the wing area is a pretty good rule of thumb, so take your wing area, let's say 450 sq.in. and punch that into the calculator and we get 63 sq.in., if we have a span of 50" divide that into 63 and we get 1.26" so a 1¼" wide aileron is what we get. Draw a line 1¼" forwards of the T.E., that will be our aileron hinge line.

That is the shape of the wing, everything else on the drawing will be structure which we shall come to in due course.

***Try different outlines, something slightly unusual can produce an eye catching model.***

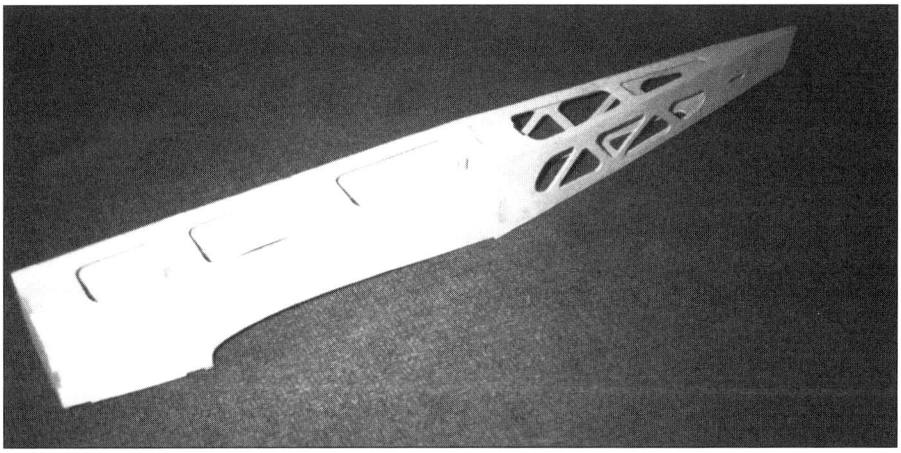

***Midwest Kits use a basic box and wing for a range of "Sort of" scale models.***

area, so it is a matter of juggling the taper into a shape that looks nice.

Slightly more sweep on the L.E. looks best and gives a sensible sort of shape to the elevators.

Don't worry too much about keeping to exact areas and percentages, models will fly with wide variations. You should now have a drawing that looks like Fig 3 although the wings and tail may be on separate pieces of paper.

## *Tailplane*

We worked out the size of the tailplane and its shape when we were drawing the fuselage side in Part III, so now it is a simple matter to lay it out.

Draw a line which will be the hinge line and drop a vertical through it which will be the centre line.

The elevators should be between 20% and 25% of the tailplane area, I prefer the larger figure because it is more efficient to move a larger area over a smaller angle.

We know the root chord and the tip chord and the

## *What have we got?*

The result of our work so far should be a drawing of the shapes of all major components of the model and we know the relationship between them. We know that the radio and tank will fit in the model and we know where the engine will fit.

The next stage is to design and draw in the structure but before we do that you will need to think about weight and structures.

***Fig 3.8 – Aerodynamics and Shape Outline***

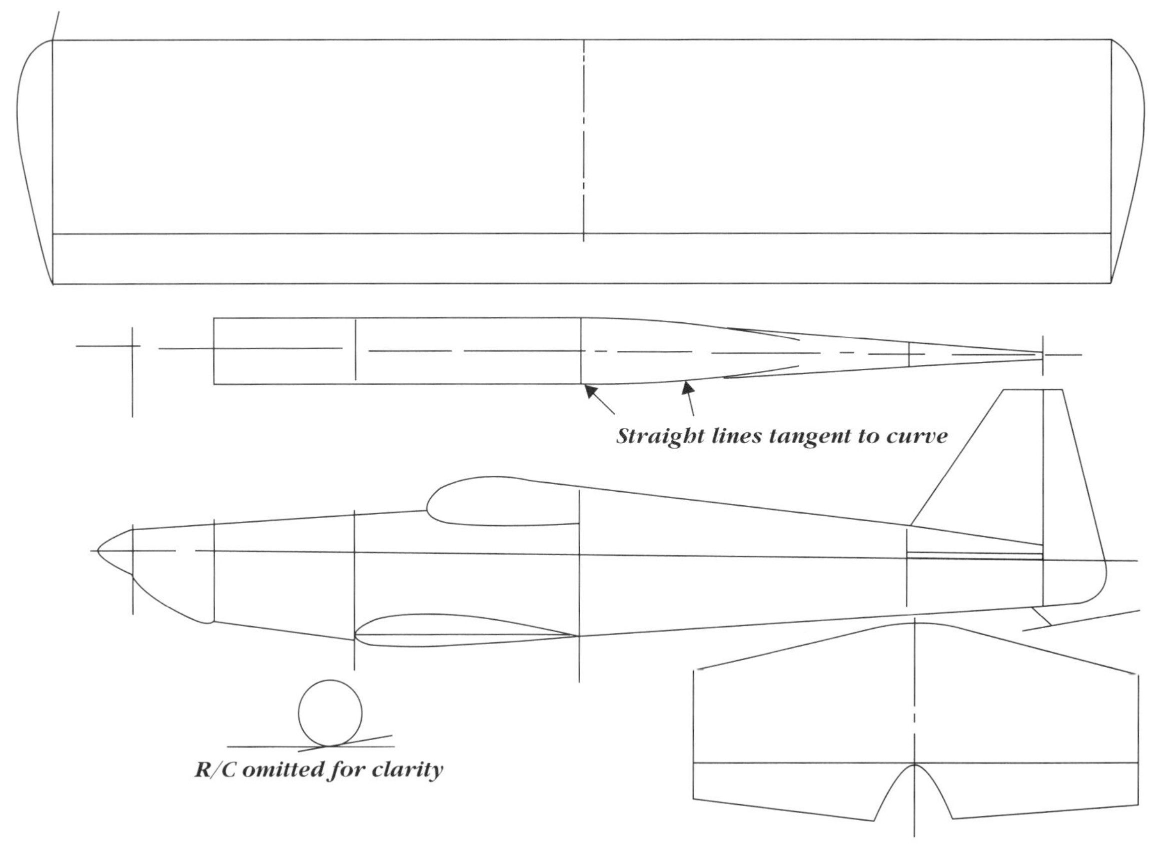

# Chapter 4

# *Lightness Equals Strength*

You have all seen those models that are built like a brick built sh . . . chicken coop. You have seen them crash and demolish themselves and perhaps you have wondered why such a strong model broke in a relatively minor crash.

I will tell you why, because the idea that large and heavy lumps of wood are stronger is wrong and in fact they contribute to the damage.

## *Inertia*

Sir Isaac Newton said that a body had a tendency to remain in a constant state unless acted on by an outside force, i.e. the apple kept falling until his head got in the way and stopped it or, a more modern example, sitting in a chair until the wife tells you to go and mow the lawn . . . this is inertia.

In the same way a model has inertia and each part of that model has inertia and when it is moving none of it wants to stop. When one part stops the rest wants to carry on and the heavier each part is, the more inertia it has.

Now imagine your nice strong, heavy model. The spinner stops because it meets the ground but everything else wants to carry on until it too, meets the ground and so your heavy wingtips hit the ground and the wing is in two (or a lot more) bits and the tail tries to get up front with the engine and your solid looking model is a heap of splinters.

Now take a lighter model, there is less inertia and and so the structure can cope with the loads.

Another effect of inertia is to make a model more difficult to fly accurately. Take a wing and hold it in the middle and rock it as though it was being put into a bank. Now add some lumps of modelling clay to each

***When one bit stops, the rest try to carry on***

***Front view of a Piper Cub fin, about ⅜" thick, bracing wires are not much thinner. Note also way fabric is used to fair fin to fuselage, something that is often missed on scale models.***

tip and repeat the action. Did you notice how much harder it is to stop the rotation once it is moving? Well that happens in flight and the same thing happens in the pitch mode when the elevators are used.

I once had a very spectacular demonstration of inertia. I was flying a .60 powered, semi-scale Fairy Firefly control line stunt model. This aircraft had wing mounted undercarriages with big, heavy wheels.

I was flying inverted and came down a bit too low and made a hard, heavy but flat inverted landing in the grass.

The model stopped in a couple of feet with only minor damage, but both undercarriage legs departed from the wings and ended up several yards further on, complete with sections of ribs and bits of spar and L.E. sheeting.

## Weight in Flight

We have seen that, for level flight, lift has to equal weight. If a model is heavier it has to fly faster or at a higher angle of attack. Increasing power is not a good option because you need four times the power to double the speed and it will not help the landing speed.

Increasing the angle of attack increases drag and slows the model until you reach a stage where the wing is flying just below the stalling point. From this it can be seen that the more weight the lower the actual flying speed.

At the same time the stalling speed will rise with the increase in weight. It is quite possible to get a stalling speed higher than the maximum flying speed of a model, this is a safe model because it can't fly but that is not the object of the exercise.

***Flying a vintage model is the best illustration of how nice light models are.***

Conversely, of course, a light model will fly faster and slower and this wide speed envelope is much safer.

A light model will accelerate and decelerate faster, this means quicker take-offs because the model will reach flying speed in less time. It also means greater safety in landings.

Consider a heavy model coming in to land in a 15 mph wind, the model has a stalling speed of 30 mph and is approaching the landing at 35 mph. Now the wind is not constant, a nominal 15 mph wind is dropping to 5 to 7 mph and rising to 20 mph.

If the wind drops to 5 mph the models airspeed is suddenly reduced to 25 mph, 5 mph below stalling speed, the heavy model cannot accelerate fast enough to make this speed up . . . instant scrap balsa.

A lighter model with a stalling speed of 20 mph can make its approach at 30 mph or less and will accelerate quickly enough to prevent a disastrous stall.

Lighter models are safer and fly better.

## Structures Full Size Classification

We can learn a lot about model structures from the full size. Most models, kits or plans, are poorly designed from the structural point of view. This could be because it is easier to manufacture the kit that way or simply due to ignorance.

Full size aircraft structures are divided into three classes. Primary, secondary and tertiary.

Primary structure takes all the main loads, the flying loads, the landing loads, everything. It is very strong. Spars, longerons, engine mountings and undercarriage mountings would all be primary structure. In some cases the skins of the aircraft may be primary structure. Any damage to the primary structure will ground that aircraft.

Secondary structure carries some loads and may have some of the loads on primary structure transferred to it. Giving examples of secondary structure is much harder, on a model we might consider leading edge sheeting as secondary but as in all cases where something lies between two extremes it is harder to define.

Tertiary structure is non load bearing, wing tips and fairings are obvious examples, very often wing ribs are tertiary structure on full size aircraft, leading edges and trailing edges are also tertiary. Damage to tertiary structure would not prevent an aircraft from flying apart from aerodynamic considerations and the possibility that the damage might extend to secondary structures.

Keep the above in mind when designing your model. After all, why make your wing tips out of rock hard ¼" sheet when some soft ⅛" with a little 1/16th can do the same job. You will save weight and make your model much nicer to fly with light tips.

Even lighter wing tips can be made by laminating strips of 1/16" sheet, these are remarkably strong but still very light. No, they won't take a full power wing tip landing but if they did the wing spar might go instead, ask yourself which is easier to repair.

***Unidentified elevator hanging on a wall, note how light all the members are and lightening holes in ribs.***

Tails can be made much lighter than many would believe. The Piper Tri-Pacer fin looks so substantial when it is covered but it consists of a strong tube which acts as hinge post and the rest is ⅜" tube and channel section, it looks positively flimsy.

To take a model example, the wild and woolly competition fun fly models have tails built up from ¼" x 3/16" and ¼" x ⅜" strip, very light and flexible, they survive some very violent aerobatics but they are prone to flutter badly.

On our simple quick built model we may prefer solid sheet for ease of construction but we can use soft sheet

***Fig 4.1 – Tapering Construction***

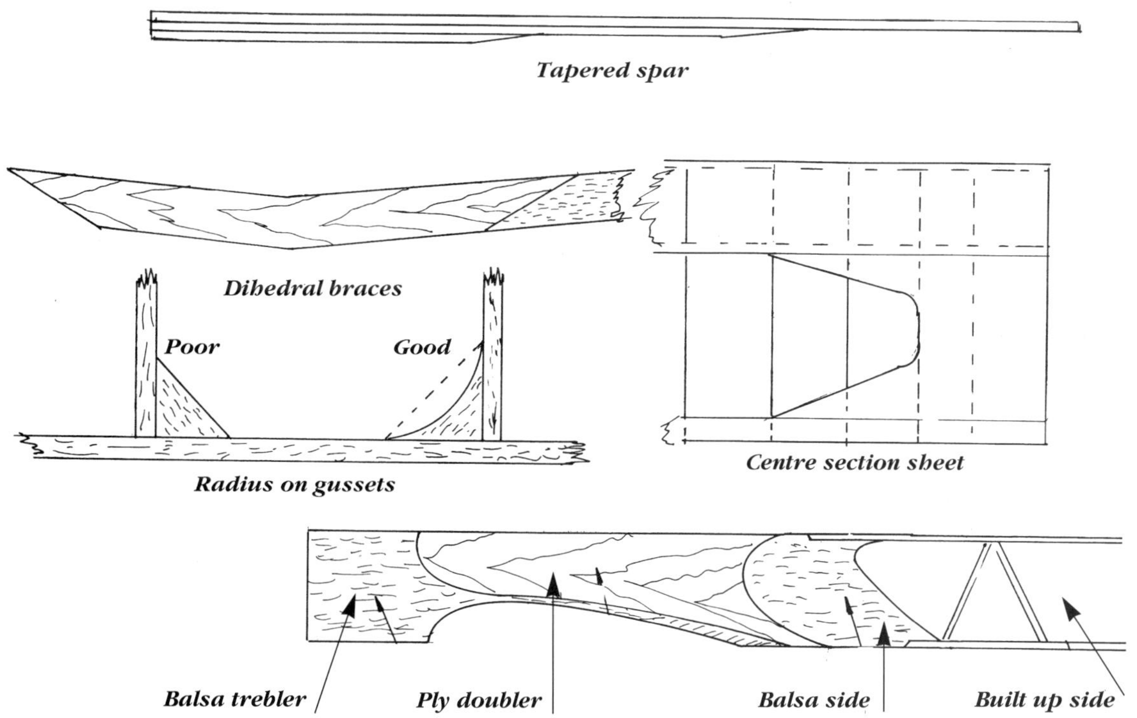

***This 15 powered aerobatic model used the tapered spar with webs and very lightweight structures, note centre section sheet blended to spars and tips. Flew like a dream.***

and punch holes in the surfaces. A strip of harder wood for the hinges will also act as a spar to stiffen things up a little.

The parts that do need to be strong are the engine mounting, the undercarriage mount, the wing attachment areas and the spars but even these can be made a bit lighter with some thought.

## *The Rotten Carrot Syndrome*

There are very few models that are designed to spread the loads properly. We see doublers that end in a straight line just at the rear of the wing, centre section sheeting that joins the L.E. sheet and T.E. at a right angle and so on.

This sudden change of section always results in a high stress area and that is where the model will break in a crash. The correct way to design any structure is to taper the change of section and to taper the strength to suit the loads.

***This Tiger Moth kit fuselage could be improved at the junction between sheet and built up section.***

***Ultra light structure on Waco PG-2 fuselage showing curved ends to doubler.***

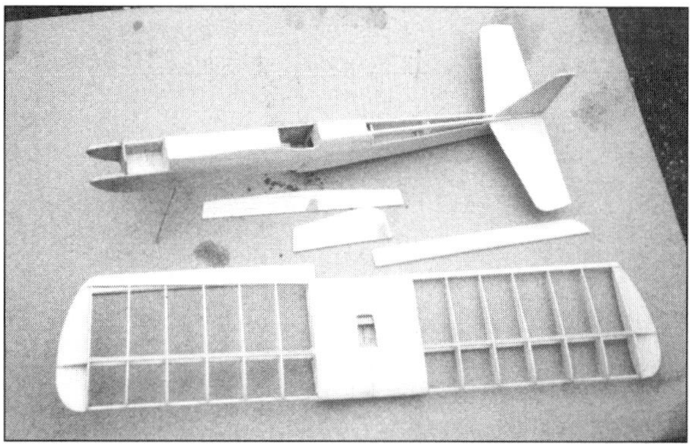

***.049 powered aerobatic model, very light, flew well, guess where the wing broke in the end.***

Look at a main spar. The normal model uses ¼" square top and bottom which may be webbed and then is sheeted over the L.E.

The strength is the same from root to tip but the load is highest at the root and lowest at the tip.

The ideal spar would be laminated from ⅛" x ¼", three laminations over the first third, two over the second third and one from there to the tip and the change of lamination should be tapered. I have only found a spar like that in one kit. Now this will not save weight but it puts the strength and weight near the root and keeps the tips light.

Fig 4.1 shows this along with other ideas for tapering strength.

The webs may be 1/16" near the root and 1/32nd over the next third with no webs at the tip.

The L.E. sheeting can be omitted and a small sub spar used to help preserve the section or the sheeting can be very soft 1/16" or even reduced to 1/32nd sheet.

Looking at the fuselage, any doublers must end in a long curve or taper and if there are cut-outs in the fuselage such as a cockpit opening and wing openings the doubler must extend beyond them. Doublers can be thinner than you may imagine, 1/32nd ply bonded to ⅛" balsa is quite strong enough for any .40 sized engine and if treblers are used right at the front it would be adequate for .60s.

If you insist on 1/16" doublers you can cut holes in them because the centre of any sheet component is under far less load than the edges.

One of my favourite fuselage sides consists of 1/16" balsa, 1/32" ply doublers and ¼" x ⅛" longerons and uprights at former locations. I have flown .46 four strokes in a fuselage with this type of construction, i.e. my Turbulent in RCMW plans service among others.

Believe it or not a slightly flexible structure will withstand crashes better than a rock solid one. Even modern full size jets have flexible wings, just look out the window and watch them going up and down in flight and twisting when reverse thrust is applied.

I have a camera plane design, the front of the fuselage is a very solid box to protect the camera and radio, the rear is built up from strip. In crashes it was found that the rigid box would break but the rear portion would be undamaged.

We don't want a floppy structure but some "give" is quite acceptable. Now read this again and digest because now we are going to draw the construction on our plan.

## *Materials*

To build structures you need materials. Yes, I know, pretty obvious but the choice of materials is important. You could double the weight of a model quite easily with the wrong choices.

There are three main types of material that we use, balsa wood, plywood and hardwood and, by careful choices, we can build the lightest and strongest possible structure.

This is where the model built from a plan scores over the one built from a kit, the quality control is in your hands and the designer who understands his materials will do the best job of all.

We will look at each type of wood in turn and see how they can be used to best advantage.

## *Balsa*

Of all woods, balsa is the most variable, this is because it grows very fast and so a very dry or particularly wet season has a much more dramatic effect on the wood.

You can be pretty certain that any plank of oak will be much like any other, but a sheet of balsa can be rock hard on one side and, in the space of two inches, go to very soft.

More often you will find less variation than that in any given sheet but any stack of sheets in the model shop will show that degree of variation between the hardest and softest sheet, and for hard and soft you can also read heavy and light.

This variation in hardness is very useful, we can use hard sheet for heavily loaded areas, medium for less stresses parts and soft for those that do not need the strength.

Not only do sheets vary in hardness, they also vary in stiffness, this is controlled in part by the hardness but also by the way it has been cut. Quarter grain sheet is reluctant to bend across it's width and so is used for things like fuselage sides and wing ribs.

The softer, easily curled sheet is best for L.E. sheeting and turtle decks where it will curve without splitting.

Strip wood should always be hard, it is used for spars, longerons and stringers, the first two have to take the main loads of the model while stringers take handling loads and soft stringers will break very easily when you pick the model up.

Solid Tail surfaces should always be made out of soft

sheet, you can't afford the weight of anything heavier at the tail end. If it is very soft you can glue hard strips round the edge or a hard spar to take the hinges and it can pay to let in a stiffener to prevent the surface warping.

As you design your model you should look at each part and decide then what grade of wood you will use for each part, consider if it will do the job in the lightest way.

Think about the tailplane, this is a large component far back from the C.G. Now assuming that you are going to make it out of sheet, which is lighter, a 3/16" sheet with harder edges and stiffener or a piece of 1/4" sheet, or would it be better to use hard strip wood? It will depend on various factors such as shape and size even the moment arm. Never forget, a surplus ounce at the tail needs up to four ounces in the nose to compensate.

Another factor that needs to be taken into account is the grain, we all know that wood splits along the grain and yet I have seen parts in kits cut-out with the grain running across the narrow dimension.

It may be the only way to get the part in and so saves the manufacturer some money but it will cost you the money to replace. At least with your own design you can get it right.

***Inside view of Taylor Monoplane fuselage showing 3/4" sq wood and ply covering. Gussets are formed by solid blocks glued into corners of joints.***

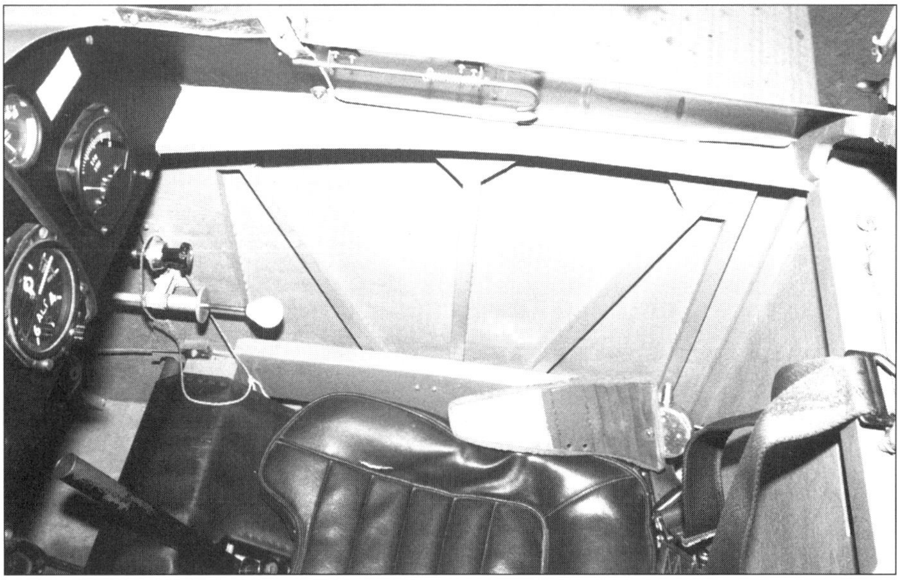

***Unidentified homebuilt fuselage under construction. Square cut-out is for wing, note large gussets spreading load in high stress area. The frame will be covered with ply later.***

## Plywood

Plywood is much stronger than balsa but it is also much heavier and so must be used with care.

There are many types of plywood and several useful thicknesses, most have their uses on our models.

The most commonly used sizes are 1/64", 1/32", 1/16", 1/8" and 1/4". The thinner sizes are often used for doublers as they tend to be rather flexible and bonding them to balsa makes a very strong, stiff laminate.

Doublers are often made out of 1/16" ply but this is thicker than you really need. 1/32" ply laminated to hard 1/8" balsa is quite strong enough to hold a powerful .40 engine, while 1/64" ply and 3/32" balsa is adequate for the nose structure of models with engines up to .15, increase the balsa to 1/8" for engines up to .20.

I once saw a model which used obechi veneer instead of ply for the doublers. The grain was running in the same direction as the balsa grain and obechi is not much stronger than balsa anyway so I formed a rather low opinion of the design.

The thicker sizes of ply are used for formers. The former on which the engine is mounted must be birch ply and, if ¼" ply is used for the larger engines, this should be multi ply with at least five layers.

There are several types of plywood, some of this is what used to be called "Double Thickness Centre Lamination", it is often made from mahogany, it is easy to recognise because the centre is thick with the outside layers being very thin. I find that it cuts badly, tends to separate and is not really strong enough.

Liteply is a much better material and can be used for some formers. Many kits use liteply for much of the construction, including all the fuselage parts. This is heavy and harder to cut-out and so we would only use it for a few parts.

Formers can be cut away leaving just an outline and the areas where other parts are attached, such as the plates for the nuts that hold the wings on.

$\frac{1}{64}$" ply is a very useful material, it feels so weak and floppy but in the right place it can be better than balsa.

My Velie Monocoupe has a fuselage built of ⅛" square spruce with an inner skin of $\frac{1}{64}$" ply in the centre area, this was powered by a Cox Queen Bee .074 and was fully aerobatic. In the final crash which demolished the top portion of the cabin, the wing, buried the engine and snapped the crankshaft, the basic fuselage was not broken.

$\frac{1}{64}$" ply can also be used for turtle decks as it will roll so easily and I have used it to sheet centre sections on wings. I also know of some models where this ply is use to sheet the L.E. of the wing.

Used carefully, $\frac{1}{64}$" ply can add valuable strength with very little weight penalty.

***Metal homebuilt, note varying size of tubes, biggest is only ¾". Turtle deck is fibreglass, rest is fabric covering.***

## *Hardwoods*

There are two main hardwoods sold for model construction. Beech for engine bearers and undercarriage mountings, and spruce which is used for spars and stringers.

***One of four bolts holding down a Cessna fin, pound coin gives scale. Note rivets holding bracket to fin. This is primary structure.***

Take the average all wood home-built. Here we have an aircraft which is built of a framework of 3/4" square and covered with 1/8" ply. There are one or two that use even thinner plywood.

Look at the size of the tube on a metal airframe with fabric covering, 3/4" tube is largest and most are even thinner. Even engine mounts use quite small diameter tubing carefully arranged in triangles.

When you look at control surfaces you can see that they are built up from a few thin pieces of wood and a spar.

Where parts are bolted together the bolts are often smaller than one might expect. To run through some examples. The Taylor Monoplane tailpane is held on by four 1/4" bolts through a couple of small metal brackets. The fin of a Cessna 152 is held down with four 1/4" bolts through four small brackets.

Perhaps an aerobatic aircraft will have more substantial bolts. No. The Top wing of the Steen Skybolt, a very aerobatic machine, is held on by two 5/16" bolts but of course the flying wires take some of the loads.

Strut braced wings on aircraft like Cessnas and Austers are held to the fuselage by two small bolts through brackets, the struts take a lot of the lift but not all.

Few designs use hardwood bearers these days, the nylon engine mount is so much easier to use.

The hardwood bearer had one big advantage, it tied the front of the model together as it ran through at least two formers. The disadvantages were that they made positioning a large fuel tank difficult and they were very difficult to replace if they got broken in a crash.

Both these disadvantages are overcome if the bearers are glued to the fuselage sides and the engine is mounted on a Paxolin plate, which will snap in a crash protecting the bearers and the engine. Being spread apart, fitting a tank between the bearers is easy.

Spruce is often used for spars and stringers and can be used for longerons. The weight penalty is low because it is used to replace hard balsa and one can use a smaller section.

Spruce is ideal for scale models where the framework is visible and the members would be over scale if balsa was used, this applies especially to stringers.

Referring again to my Velie Monocoupe, the rear portion of the fuselage was made from 1/8" square spruce longerons with 1/8" square balsa cross members, the crash did not damage that part of the model either.

An incredibly strong fuselage can be built with spruce if you gusset each joint with 1/64" ply. I actually built a fuselage frame with hard 3/16" balsa and gusseted each joint. I decided to scrap the fuselage and I used it like a club against a metal chair and the bench, holding it by the tail end and smashing it down. I had to hit it against the bench and the chair as hard as I could several times before it even started to break up.

## Strength and Lightness in Full Size Aircraft

I have mentioned using full size classification of structure but let's look at some examples of full size construction. You will be surprised as just how lightly built most aircraft are.

## The Application of Models

Models do suffer stresses that full sized aircraft could not survive. We pull "G"s that would have a pilot squashed flat on the cockpit floor and of course many, if not most, of our landings should be called controlled crashes, but we can still make our models a bit lighter.

Even allowing for these higher loads we could use smaller sizes of materials that we normally do.

My Monocoupe 110 Special has the wings held on with four 4mm nylon bolts, I could have used 8 BA steel and still have been completely safe. I had a .40 powered control line Pitts special once, and that used two 6 BA bolts to hold the top wing on.

Take the Taylor Monoplane, a favourite subject because I spent a lot of time doing the scale drawings of the full size aircraft. A 1/4 scale version would only be 63" span and the fuselage frame could be 3/16 square hard balsa and 1/64" ply or 1/16" hard balsa skin for a very strong scale structure, such a model would be quite happy on .40.

By thinning about weight and strength, using materials to the best advantage it is possible to build a light model that will withstand appalling punishment and only need minor repairs.

# Chapter 5

# Fuselages

Many people draw their plans with single lines to represent parts, for example just the front of the spar and the rear of the T.E. They also use single lines for the formers and fuselage sides.

This is bad policy in my view for two reasons. The most immediate reason is that you have to draw all the parts using previously drawn items. You need the fuselage side and plan views to draw the formers. It is all too easy to get mixed up when taking these measurements if you only have one line.

Imagine that you are only using a single line for each former. The front former would have to be represented by the front face because that is where the engine sits, but the next one back would have to be shown by the rear face which is where the wing L.E. locates and the former at the T.E. would again have to be shown by the front face . . . See how easy it would be to get confused?

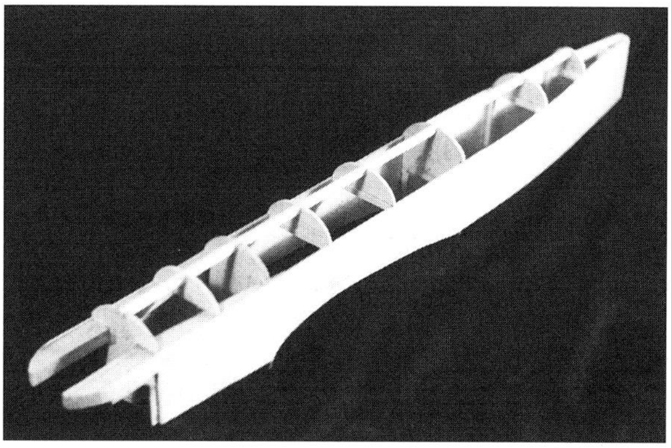

***The majority of models use the basic box fuselage . . . this is Miss 57 from the RCMW Plans service . . .***

***Fig 5.1 – Construction Layout***

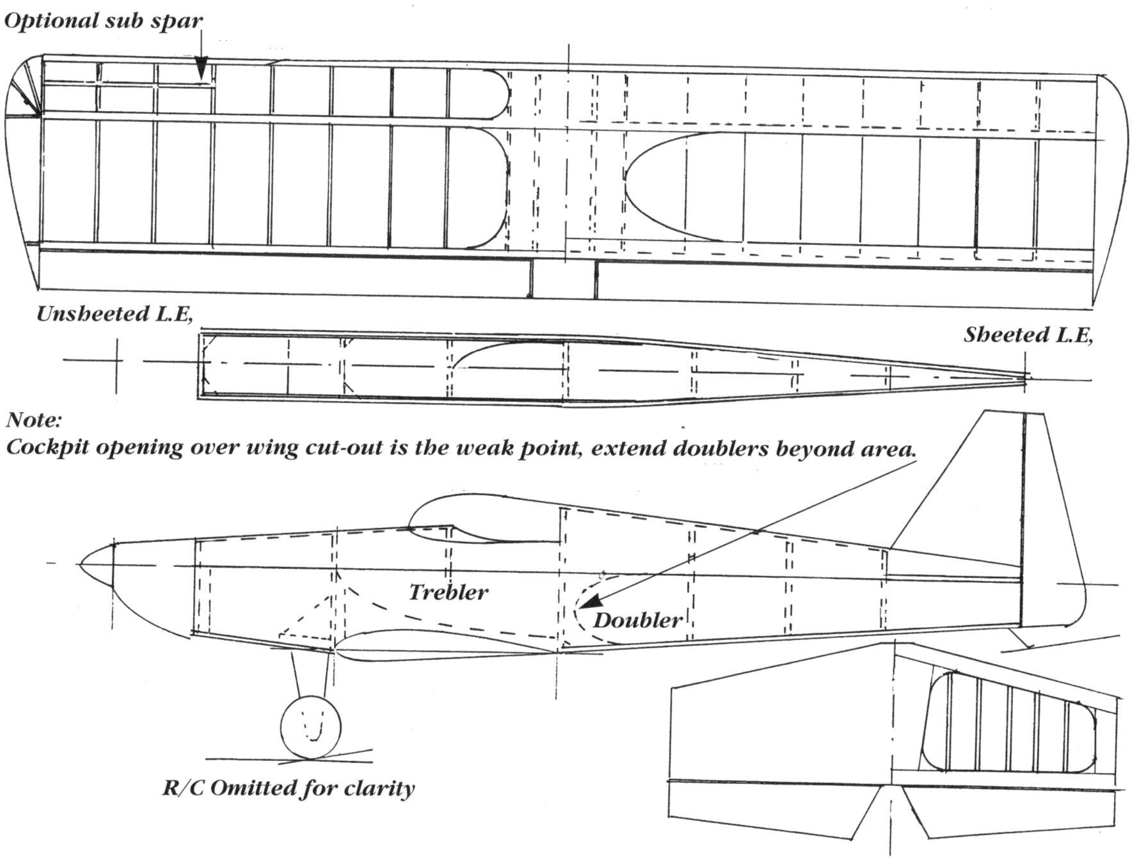

***... some just leave it like that ... Blue Movie 2 from the RCMW Plans Service, there is not much else you can do with Correx ...***

The second reason is that you just might want to offer your plan to a magazine and they will need more than a few thin lines. A typical plan is shown in the drawing Fig 5.

## *Fuselage*

In this part I shall tell you how to lay out your drawing but I won't cover everything in great detail because anyone designing a model should have built one or two kits and so will have a little experience. For example, you shouldn't need me to tell you that the engine mounting former must be plywood and must be firmly attached to the sides. Designers use many standard techniques so, if something is not covered in detail, use a method that you have seen elsewhere.

***... some dress it up a bit ... Trivial Pursuit shows that angular turtle deck can look smart ...***

The side view will show any doublers and the formers. Indicate the doubler, not forgetting to make it end in a curved line away from any other changes in section (You did read the last chapter, didn't you).

Formers come at logical points, the engine mounting bulkhead, at the L.E. and T.E. of the wing and at the L.E. of the tail plane. You will need at least one between the wing and the tail and the turtle deck may need a couple of extra formers.

Drop lines to the plan view and draw the doubler and formers in to this. If you are feeling lazy you can indicate the thickness of the sides and doublers in the plan view at the former locations only.

Now to plot the formers. Draw a long datum line and drop verticals from this, also extend the verticals upwards. It is at this point that square protractor comes into its own.

Now, using a pair of dividers take the distance from the datum on the side view to the bottom at the engine former and transfer this to the first former, repeat with the top dimension and then using the plan view take the width from the centre line to the side and transfer this to each side of the former drawing.

Using your square protractor plot out the square shape of the basic former. The top portion will be plotted later. Repeat for all the other formers.

There are various styles of turtle deck, you can leave it flat (Ugh!), you can make it rounded or a sharp angled deck.

The sharp angled top is easier to plot as you can use fixed points on each former. The rounded turtle deck looks best and can be plotted using the height and width and drawing it with compasses. A picture being worth a thousand words you will find two drawings showing the method elsewhere in this chapter. Fig 2.

The method shown for plotting round topped formers works well, but sometimes you may want a different shape to the turtle deck and you will find that intermediate formers cannot be plotted easily. The simple way is to leave them oversize and sand down with the aid of a straight edge until they blend in.

Curved turtle decks will need to be planked and this method of construction allows you to use formers that vary quite dramatically. This can make for very attractive shapes which are not often seen these days.

When drawing formers out don't forget to allow for the thickness of the sides and top, and remember that the doublers do not extend to the tail, all these are good reasons for drawing the model in full.

Other items that have to be added to the fuselage drawing would be wing bolt plates, any reinforcement for attaching the undercarriage, perhaps a sheet insert for gluing down the tailplane and a ply mount for the tail wheel and skid.

Draw all these parts in on both drawings. Very often you will spot possible problems while doing the drawing which is a lot better than finding them when construction is well under way.

For our initial design we can use a very simple undercarriage, the bent dural U/C bolted to the bottom of the fuselage so incorporate a ply plate just in front of the wing and use triangular stock to reinforce the joints between the plate and the sides and also between the engine mounting former and sides.

This whole area is primary structure as defined in Chapter 4.

*Fig 5.2 – Former Plotting*

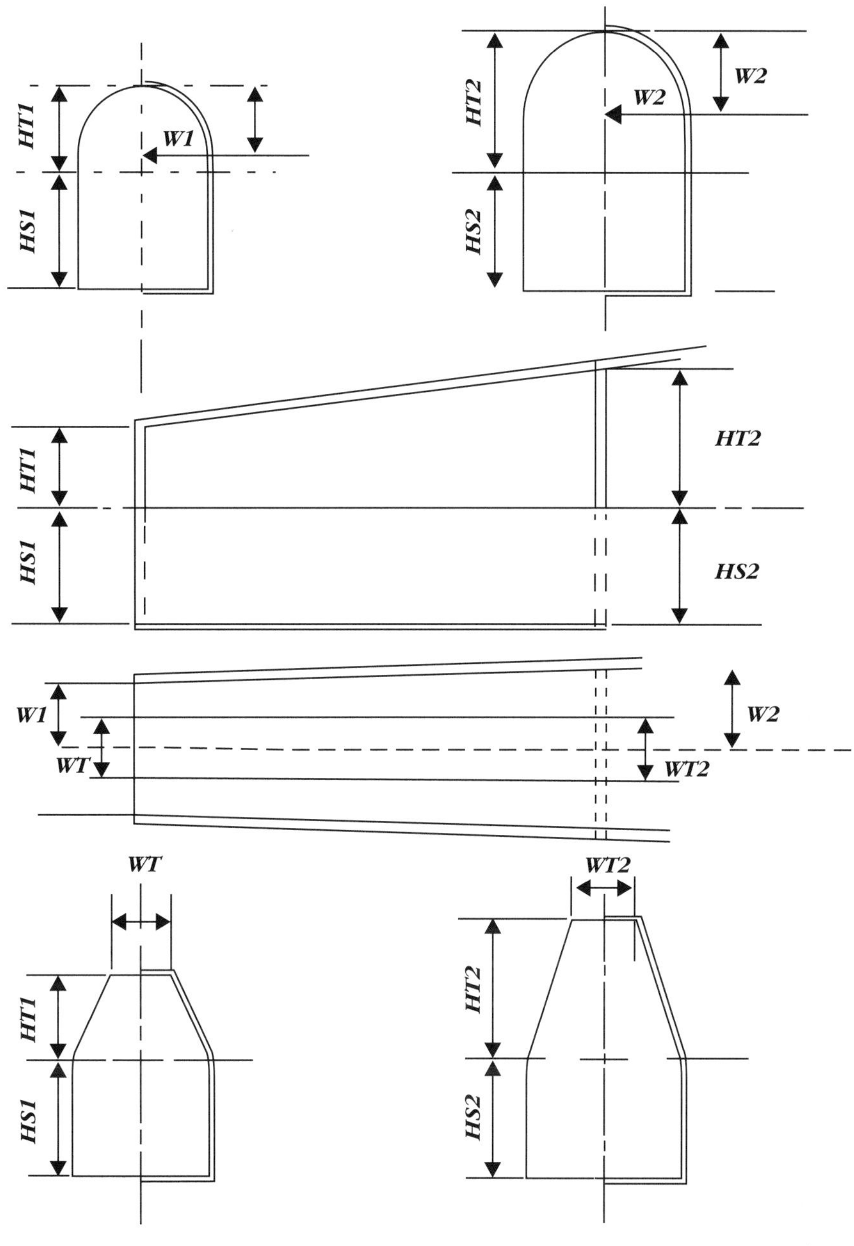


## *Advanced Fuselages*

The design of your average sports model is pretty basic and there are limits to what one can do with a plank wing and modified box fuselage with dural U/C.

Add round fuselages and tapered wings to the permutations, throw in a variety of undercarriage types and you can produce some really different looking models.

You may feel that such complications are not worth the effort and for sports models this could be true, but when it comes to scale models some of the nicest subjects are more complex.

The aerodynamics remain the same, the problems come in designing the structure so that it is still light and relatively easy to build.

Fig 5.3 – Rounded Box

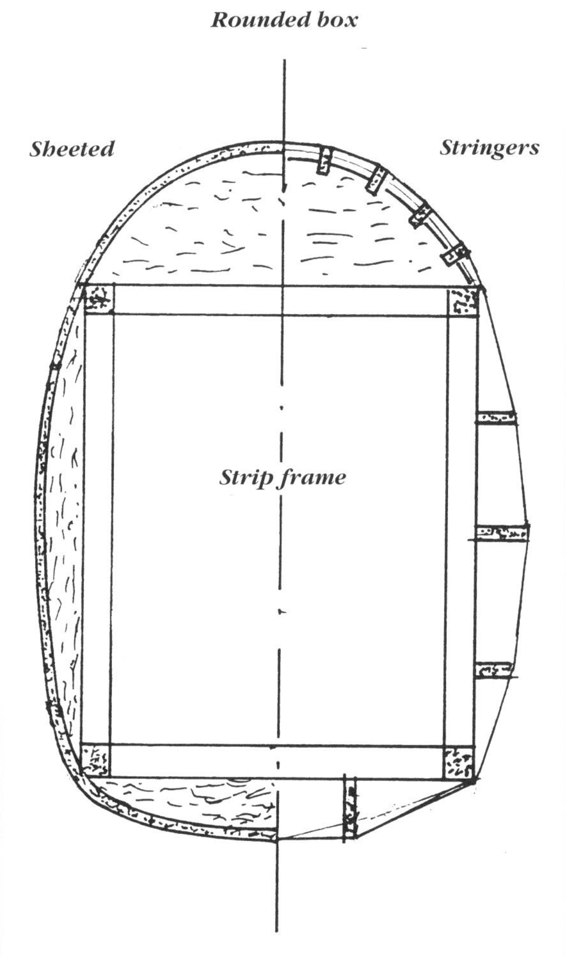


## Rounded Fuselages

If you study 3 views and photos of full size aircraft you will see that all fuselages are variations on a few methods.

The slab sided fuselage has been dealt with but a few, like the Piper Cub have a stringer or two down the side which just gives a little more shape.

More rotund fabric covered fuselages will be the simple box with sub formers and stringers. For this type of fuselage it is well worth building the box up from strip like a vintage model, as the stringers will add strength and yet the whole thing will be no heavier than a plain box. Fig 3.

The next type is the metal covered fuselage and here careful analysis can pay off. Let's take two examples, the B-17 Flying Fortress and the Ryan STA or PT series.

The first C/L B-17 that I built was planked, a very long and tedious task but then I found better drawings and studied more photos and I realised that the fuselage is a cylinder and a cone, the only double curvature was at the nose. You can do the whole thing with rolled sheet. This is what I did with my second B-17 and the nose sheeting was slit and curved and pulled in just like short planks.

Look at the Ryan STA and you will see that it is the same, straight lines which are very easy with balsa.

It is possible to get some double curvature into balsa sheet. The method is quite simple. You need a frame work of formers with substantial stringers about 3" apart.

Take soft balsa sheet and soak it in water for a while, if you add a little ammonia the soaking time is reduced to about 15 minutes.

Now take the sheet and carefully bind it to the framework with bandage and leave it to dry. Two opposite panels can be done at a time. Once the panels have dried they can be trimmed to fit and glued into place.

When building a fuselage like the B-17 it would be a waste of weight to have much internal structure, as the skin and ¼ sq. longerons would be quite strong enough, but some sort of jig is needed to build the fuselage.

The simple answer is to buy a piece of wood about ⅜" (10mm) by 3" or 4" and cut slots to take the formers which can be lined up and spot glued in place. Once most of the fuselage is built it can be cracked away from the jig. Fig 4.

This type of jig can be used on other models, the Hirth Acrostar has a fuselage that goes from slab sided to round over a very short distance and the only way to build it is with the jig described.

One of the most beautiful series of aircraft were the Lockheed Vega, Sirius, Altair, etc.

***Internal view of rear fuselage of an Auster undergoing restoration. Thin outline ply formers support stringers, nearest former is coaming which takes cabin glazing. Main framework is steel tube.***

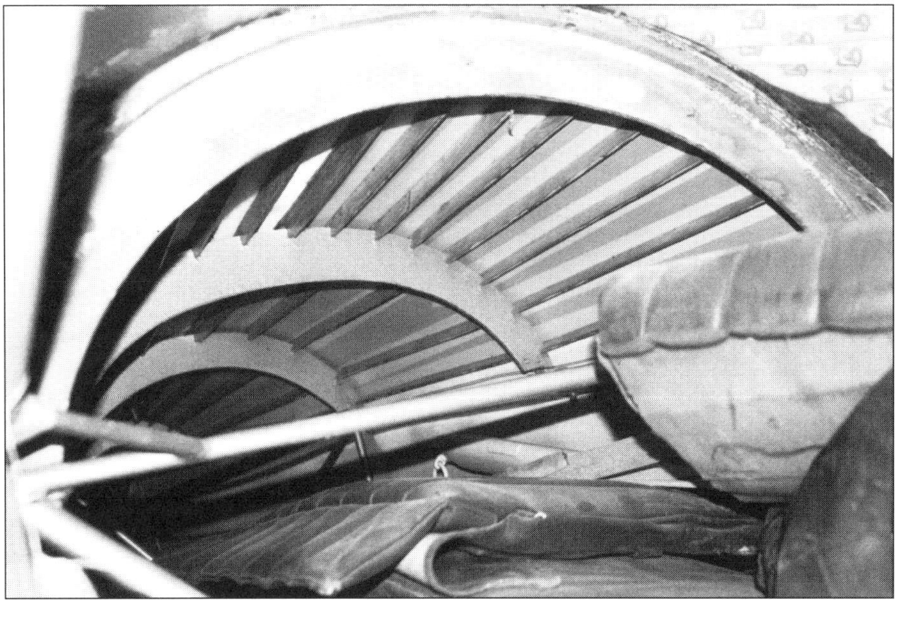

This range of aircraft used one monocoque ply fuselage shell formed in a concrete mould and a standard wing with slight variations on dihedral.

From these components Lockheed built the parasol wing Air Express, the shoulder wing Vega, the low wing Altair, Sirius and Orion, some with and some without retracting undercarriages.

These aircraft were very famous, Winnie Mae for example held the record for a round the world flight.

I have often wondered why some kit manufacturer has not produced a series of models of these aircraft. Even though I dislike fibreglass and foam I can see the advantages here. One main GRP moulding and a few small ones, one basic foam wing and a variety of sheet tail surfaces and he would have five different kits.

From our point of view the Lockheed series would need to have the fuselages planked, which is the reason that I have never built a model of one of these aircraft.

***... with stringers on the sides of the box ... like Tequila Sunrise ...***

## *Meanwhile, Back at the Drawing Board*

The average sports model has a hard life (and a short one sometimes), and so we may not want to go to such lengths as described above but we can use the techniques.

***... while others hide the fact completely ... Pee Tee Sportster is built as shown on figure 5.3***

*This Auster has very boxy fuselage but three stringers add a little shape.*

*. . . but a few do not use the box at all. B-17 built as shown in figure 5.4 . . .*

*. . . a few use a box at the front . . . like the Hirth Acrostar.*

*You can make some really attractive shapes*

The slab sided sports model can be transformed by adding stringers down each side, instant round fuselage.

A nice double curvature turtle deck can be planked (NO! The man who said hollow block, it's heavy and very expensive).

Combinations of all the methods can be used to produce some really beautiful fuselage shapes. The photos and drawings will give you some ideas. Just remember the rules, keep it light, no sudden change of section, spread any load and tertiary structure does not carry any loads.

*Fig 5.4 – Simple Fuselage Jig*

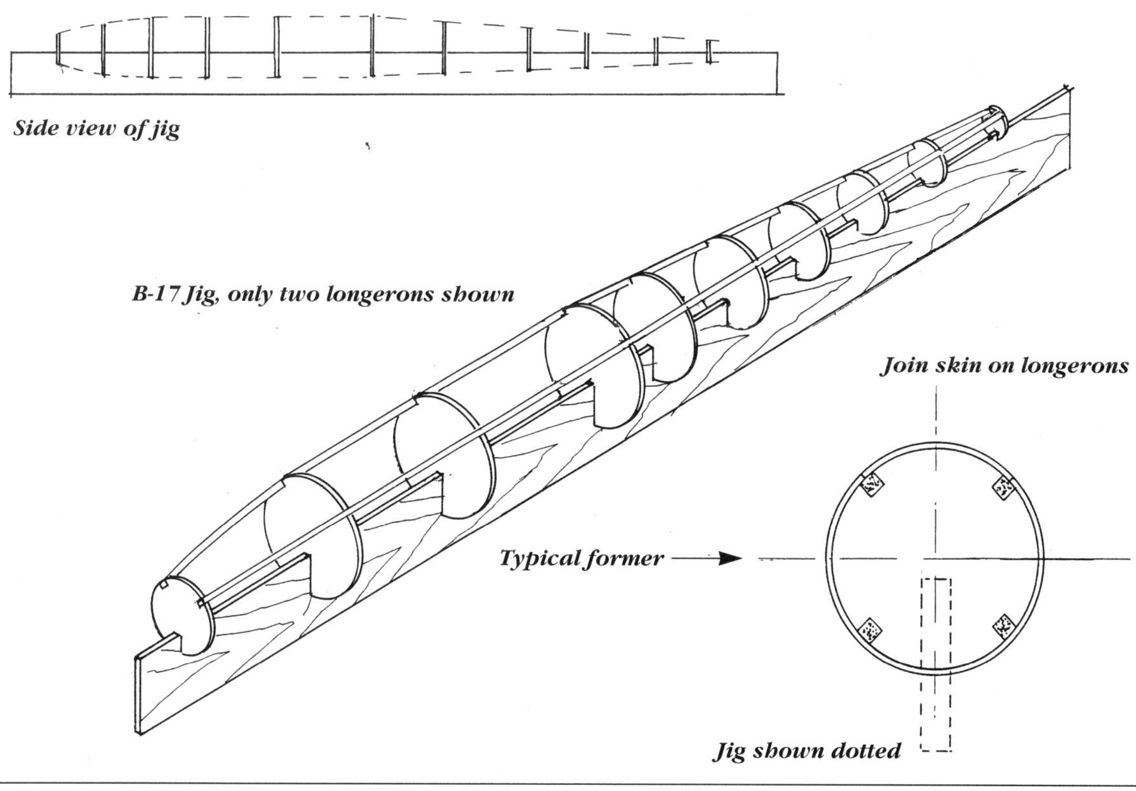

# Chapter 6

# Wings and Tails

## Wings

The wings are considerably easier to draw out, the cross-section is already decided by the airfoil section so it is just a matter of deciding on the size and position of spars etc.

On a model the normal location for the main spar is at the thickest point in the section, around ⅓rd back from the L.E. On larger wings you may want to put another spar at ⅔rds chord.

We have talked about tapering the spars and while this is the ideal I predict that few people will bother, preferring the standard type of spar, i.e. ¼" square top and bottom possibly with webs.

Leading edge sheet does add strength and provides a smooth surface over the most important section of the wing but it is not essential on smaller and lighter models such as powered gliders. Nose ribs between the full ribs or smaller sub spars can give equally good results and are cheaper and lighter.

For the more highly loaded and stressed models the 'D' box is much stronger and will prevent warps and

***The simplest wing of all, parallel chord, no ailerons and no sheet, suitable for very small light models. This is American Dragon for .8 diesels but has been scaled up to 60" span for .40 four stroke.***

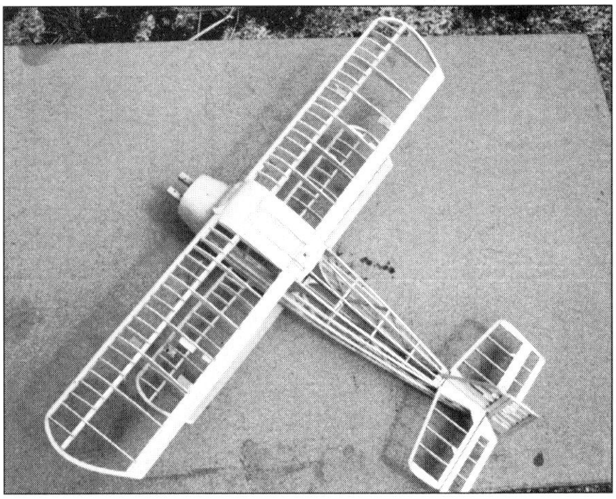

***Fig 6.1 – Wing Sketch Page***

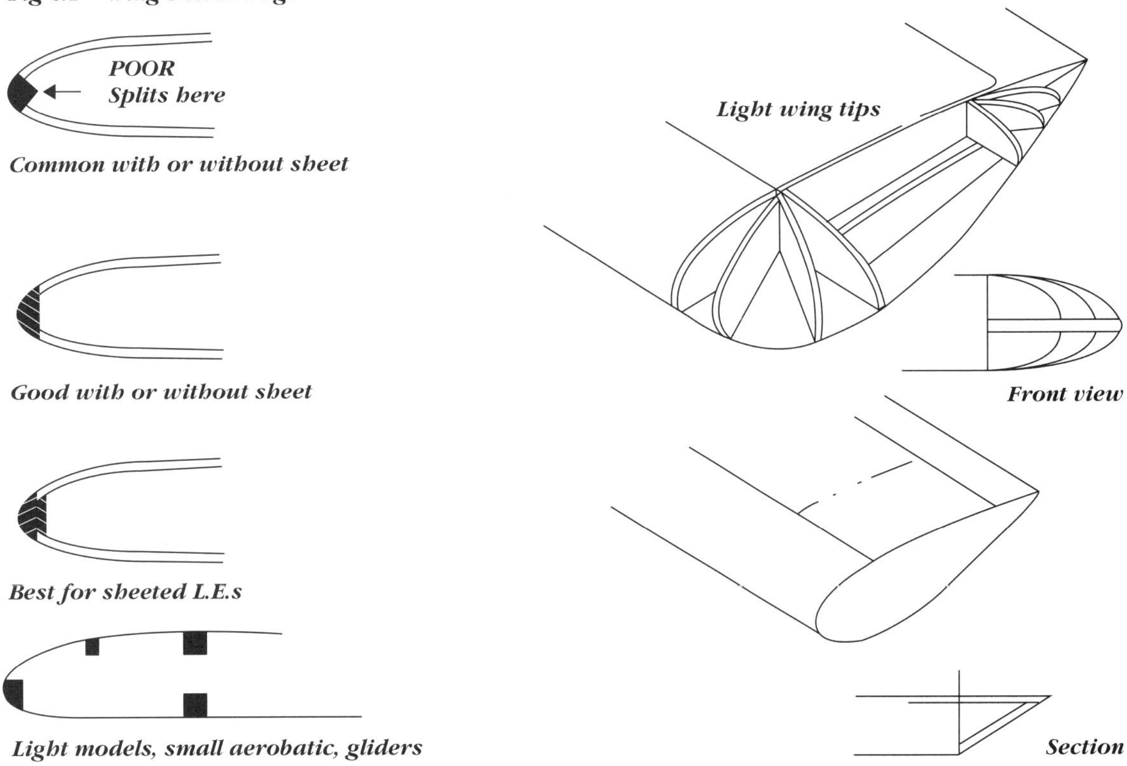

***Elliptical tips on parallel chord look nice, inset ailerons are very effective on this Monocoupe 110.***

distortion of the wing. Be warned that it is almost impossible to remove a built in warp from this type of wing.

Rib spacing will depend on the type of model, the size and loads. Closer ribs at the centre especially if a wing mounted undercarriage is used and wider spacing out beyond the centre section sheeting.

All sheeting should follow the rules about tapering off the strength, large radii at all corners or centre section sheet tapered into the L.E. and T.E. over two rib bays will prevent that sudden change of section that snaps so easily.

Trailing edges can be made with a strip of ¼" sq. and two pieces of 1/16" sheet or of more substantial strip without the sheet. The T.E. does have to take the loads of the aileron hinges and so must be strong enough for the task.

Leading edges do not need to be very strong. One very common type of L.E. is ¼ sq set on one corner, this is not good because a slight tap will split the ribs. A better L.E. is sheet glued flat to the front of the ribs.

My favourite L.E. is a variation of the above when used with L.E. sheeting. The L.E. is ⅛" sheet, it can be quite soft, this is glued to the ribs and sanded to blend in perfectly. The sheet is glued on and trimmed back flush with the front of the L.E. and then a capstrip of ⅛" sheet is glued on the front and sanded to the correct section. This system is very strong and very easy to make and always looks very neat.

Joining two wings can be done in various ways. Some American kits have used glass tape and resin just like joining foam wings, it works if you don't mind the mess and effort of getting the joint smooth.

Small models can just have the wings butt joined and rely on good glue joints and the covering material, but this is only for models with engines up to .09.

The normal method is to use dihedral braces on the spars and L.E. and T.E. This system has worked for as long as models have been built, just remember to taper the braces or the wing will snap where they end.

Strip ailerons are normally solid, quite often ready shaped T.E. stock can be used but do not compromise the airfoil section by using the wrong size of wood, better by far to shape your own stock from thicker wood.

Tips must be kept light, remember inertia? A couple of suggested methods are shown in the drawings.

## Dihedral

Dihedral is used to provide lateral stability, on a model with no ailerons it is essential but not on others.

A model with no ailerons needs about 5° or 6° on each panel, this will give reasonable rudder control, less dihedral will reduce the efficiency of the rudder.

***Fig 6.2 – Wing Shapes***

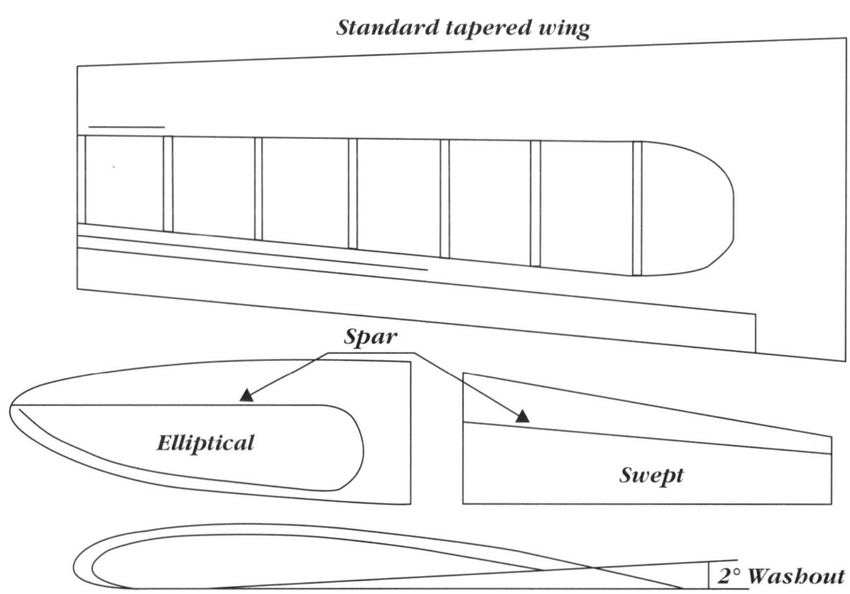

***Tail with laminated outline is very strong and light, this one is Trivial Pursuit for 1.5cc engines, spars could be added for larger models.***

they do add some stress areas in the wing, however, they are essential for scale models. I know that there is a so called scale Piper Cub kit that uses strip ailerons but you are not going to sink as low as that when you design a scale model are you?

The main things to remember when designing inset ailerons is that it should be 1½" to 2 times the width of a strip aileron and that you should use gussets to spread the loads and remove the stress areas in the corners.

Fig 6.3 shows a typical inset aileron. The aileron bellcrank can be reversed and the horn can be fitted on top of the aileron if desired. Note that the bellcranks point in opposite directions in each wing so that one is pulling as the other pushes.

Another snag with strip ailerons is that they can twist, this will reduce the angle out at the tip where you need the deflection most. You can use much harder wood which will resist this twisting but that is needless weight just where we don't want it.

One way to overcome this is to connect the controls to the centre of the aileron by means of a system of bellcranks and pushrods or, alternatively by using a servo out in each wing. This allows the aileron to be used as flaps via a mixer, probably the only aerodynamically valid reason for using strip ailerons.

Inset ailerons are much more complicated to make and

It is desirable but not essential to have differential aileron movement, this means that the up going aileron should move up farther than the down going one. This is because the down going aileron creates more lift, more lift equals more drag, more drag tries to pull the aircraft towards the down aileron which is the exact opposite of what we want.

By making the up going aileron move further we increase the drag so that it is equal to or more than the down aileron. The Taylor monoplane homebuilt, specifies aileron travel as "25° up and almost no down movement."

This differential has another advantage. The down going aileron in effect increases the angle of attack of the wing in that area, at slow speeds this can initiate a tip stall on that side, reducing the throw with differential helps to delay the tip stall. Fig 6.4 shows how to set up the aileron bell crank for differential, Chart Micro Mold make an adjustable bellcrank which is useful for this.

***Tail assembly using built up and core methods. Core systems is very good for scale control surfaces.***

## *Tail Surface*

For initial design exercises the sheet tail is the best way to go but you can build them up with something like ¼" x ½" outlines and ⅛" x ¼" strip ribs.

Used medium sheet for the solid tails and hard stock for built up tailplanes.

Of course the drawing for solid sheet tails is done as soon as the outlines have been drawn and I hope that I don't have to describe how to draw the built up system.

It is rare for a model to finish up nose heavy and so anything that can be done to keep the tail light will help, don't forget that an ounce of surplus weight at the back end will need three to four ounces up the front, this is useless weight which we need like a hole in the fuel tank.

Tail surfaces can be built up, there are three common methods. Ordinary straight strips of wood with ribs as shown in drawing 6.4 of this chapter.

More curvaceous shapes can use the same method but the outline can be done with laminations as described for wing tips.

Another method which is very good and popular is the core method, a thin sheet core is cut to the shape of the tail and then ribs and spars etc are glued on each side, when covered it looks just like a fabric covered tail but is very strong and also light. Combining this method with the laminated outline makes a wonderfully strong light structure.

Weight can be saved over a solid tail by building the structure from strip and covering it with 1/32nd balsa, this makes what is, in effect, a box structure with webs which is incredibly stiff for its weight, but you must use soft wood throughout.

One method that was very popular some years ago was to build an outline with diagonal bracing (Warren Girder) and then put a spar top and bottom and fill in the ribs with small triangular shaped pieces. This method was used on the aerobatic models of the day and is light and strong. The tail needs to be stiff enough to prevent flutter. The top Competition Fun Fly models use incredibly light tail surfaces with very large control surfaces but these models are not designed to be flown fast and most warn the builder that flying straight and level at full throttle will cause the model to flutter and self destruct. I have seen it happen several times.

***Auster tail assembly. Note how fabric on fin is used to make a fairing to fuselage, typical of most fabric covered aircraft.***

***Fig 6.5 – Tailplanes***

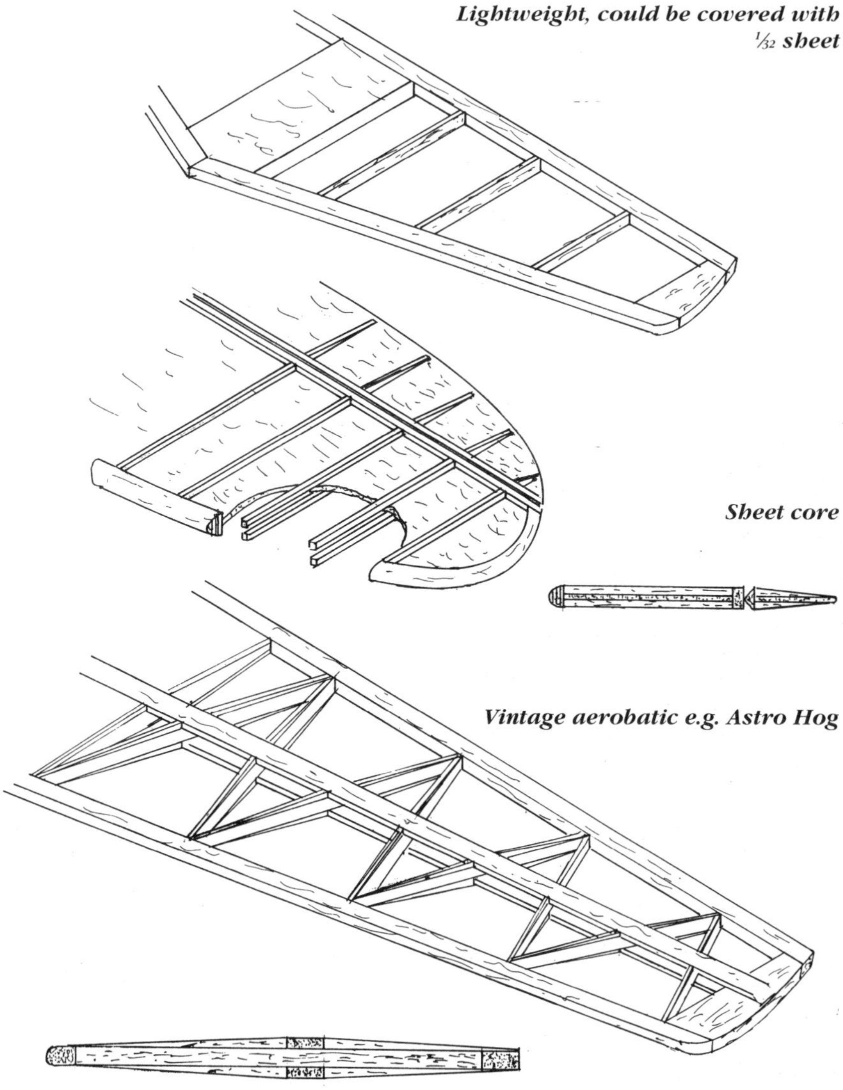

# Chapter 7

# Undercarriages and Other Bits

The main parts of the model are the fuselage and the wing but the design of the undercarriage and its mounting is of major importance, this area is primary structure.

The cowling is also important because a badly designed cowling can cause the engine to overheat which will damage the engine and may cause a crash.

## Undercarriage

The undercarriage is only used for a short time at take-off or landing, however, it is a vital piece of equipment. It stops the propeller being worn down and the belly of your model being scraped along the ground so it must work properly and not collapse.

To work well the wheels must be in the proper place or take-offs will be difficult or downright impossible. The undercarriage must also be of the right type.

Trying to use a tricycle undercarriage on anything other than short grass or tarmac is a dead loss, the long grass will hold the nose down and even fold the nose leg back enough to allow the prop to get into the grass, slowing the engine. If the main wheels are too far back not even the most powerful elevators will get the nose up . . . No take-off.

The main wheels on a trike should be just behind the CG, so close that if the model is rocked back onto its tail it will stay there. In this way the elevator will be able to lift the nose and allow the model to take-off.

The nose leg takes an awful pounding so it must be

***Simple dural U/C can be held on with elastic bands or . . .***

*Fig 7.1 – U/C location*

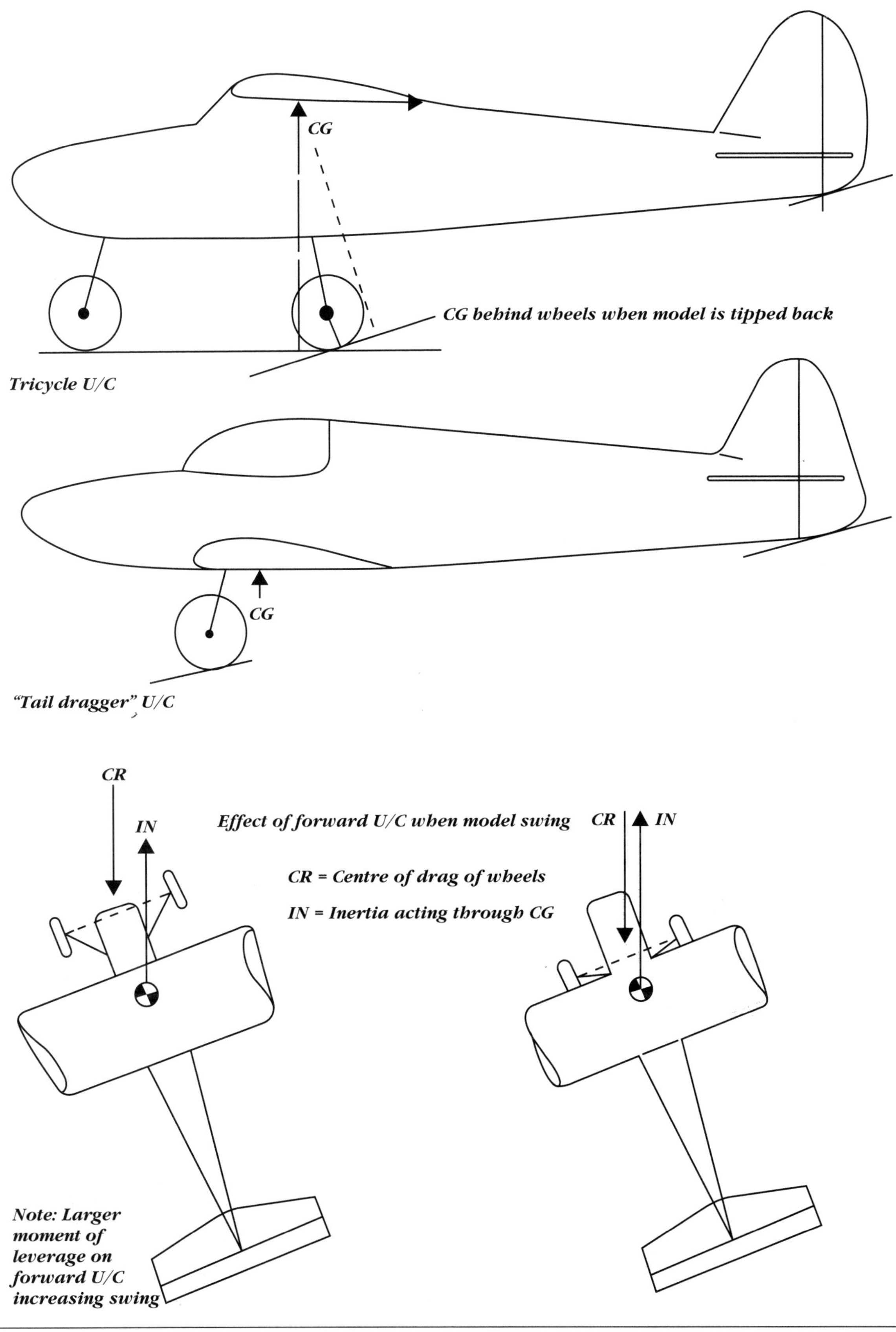

***. . . bolted on with nylon bolts. (Number plate bolts from Halfords are good).***

strong enough and it must be mounted firmly, not only that but the former must be solidly attached to the fuselage.

There is a certain kit around which has a trike U/C, it also has a tank hatch on top of the nose. I have seen several of these kits and in quite gentle landings the leverage of the nose leg has pulled the front former out, because the former was only held at the sides and not across the top.

The owner of one of these models drilled two ⅜" holes in the top of the front former and F-2 and epoxied hardwood dowels between the formers to try and tie the front former in place. It worked.

Tail dragger U/Cs need to be positioned with care. The ideal location is just under the L.E. of the wing. Slightly further forward is better on grass but do not be tempted to put it too far forward for that is the way to ground loops.

Mounting the U/C can be on the fuselage or the wing. For sports models I prefer fuselage mounting because it is lighter, it can be mounted on the ply former or bolted to a small plate in the bottom of the fuselage.

The wing mounted U/C needs hardwood blocks and ply doublers on the ribs which are heavier but it can look better and does allow for torsion springing. On low wing trikes it is the only option in most cases.

The undercarriage takes tremendous loads, just try twisting a Dural or 10 SWG undercarriage leg back into shape and see the effort that has to be used. A lot of the landing loads have been absorbed by that twisting but much of the load has been transmitted into the fuselage or wing. That load has to be spread.

Look at the drawing of the wing mounted undercarriage, note how the doublers on the ribs spread the load and transmit it to the main spars. Note that the torsion springing will absorb some of the loads but not all.

On the fuselage mounting note that the mount is braced with triangular stock to spread the load in the joint and that the stresses are then transmitted to the ply doublers. One of the best methods of mounting a dural U/C on a trainer is to attach it with elastic bands, if the landing is too hard the U/C will flick backwards and the model will settle on its belly with minimum damage.

***Extra cosmetic struts dress up a normal single leg wire U/C.***

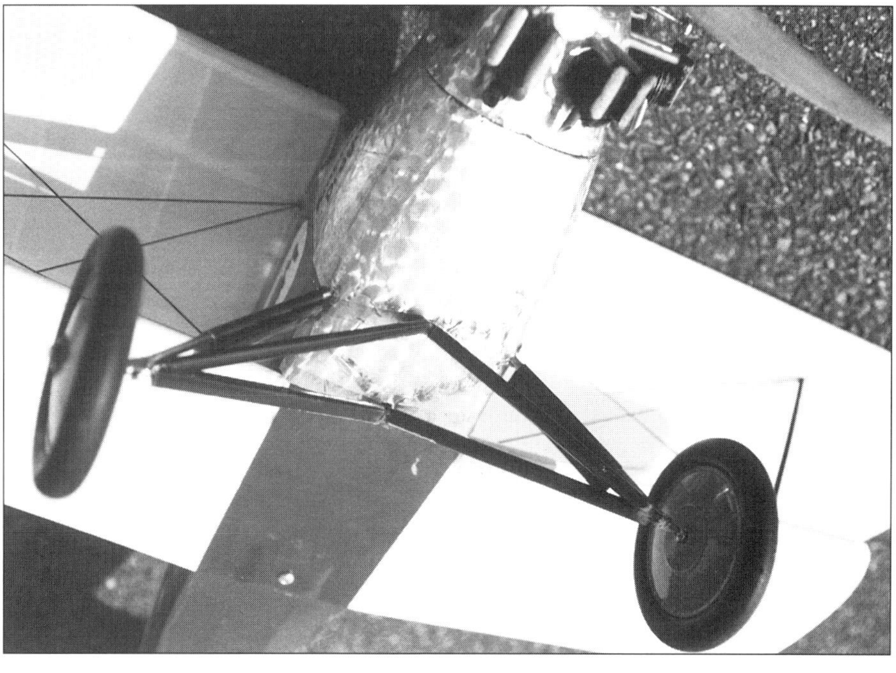

*Fig 7.2 – Undercarriage Sketch Page*

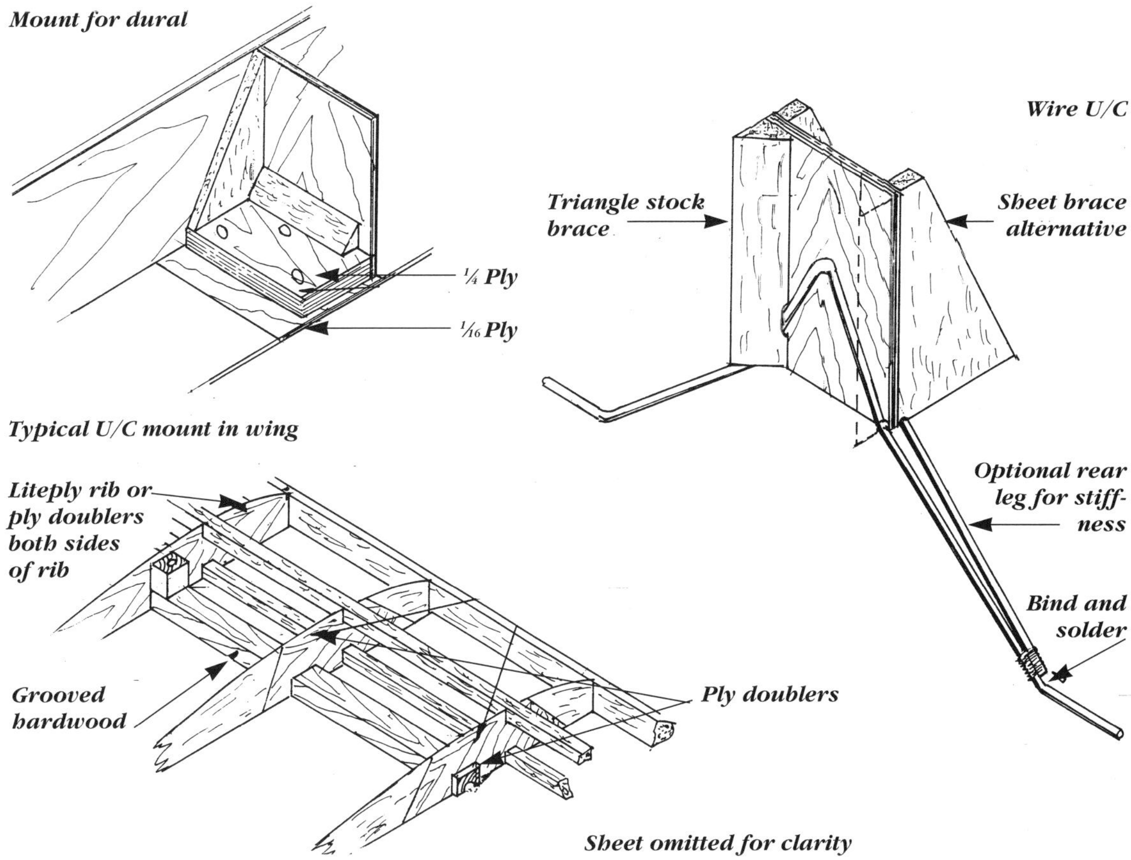


## Cowlings

Cowlings may not be so important from the structural point of view as they can be classed as tertiary structure, but they can have a very serious effect on the engine cooling if not designed properly and it is very frustrating if you cannot get in easily on the field to sort out some minor problem.

Cowling the engine neatly is often a problem, many models just use an exposed engine with just the crankcase covered by a cowl that is part of the fuselage.

***This engine is normally cowled, note how baffles take air from top of cylinders downwards, forcing it through the fins.***

This system has many advantages as almost every vital requirement is covered.

The vital requirements are, good cooling, good access to the engine and controls and security of attachment.

Good cooling needs a flow of air directed over the cylinder fins and good outlet for the hot air, this should be bigger than the inlet. Many people think that a nice big radial cowl is perfect, but in fact the vast intake and limited outlet makes for poor cooling.

The answer is to fit a plate with dummy engine in the cowl and this forces air round the cylinder. The plate has no serious effect on performance as proved by my Monocoupe 110, this model was 44" span with a cowl that hid the O.S. 15 FP engine, over 4" in diameter. One might expect the plate blanking the cowl off to act as an airbrake, not a bit of it, that model was very, very fast in flight just like the full size.

Control line team racers and speed models are operating at the very limits of engine performance, good cooling is essential and yet, if you look at the cowls on these models they have tiny air inlets. Of course they are flying very fast so there is a lot of ram air but the main reason is that that air is forced through the fins and right round the cylinder which provides the most efficient cooling.

Engines in large cowlings would benefit from some baffles to guide the airflow round the cylinder and through the fins. Full size engines use these baffles, the baffling on a twin row radial like the Bristol Hercules is something to behold.

*The inside of a well designed cowl, note shape of duct round cylinder and crankcase cooling scoop in other applecheek. Note also doubled wire for U/C, just about the best wire U/C going. This is Moondancer.*

*How radial cowls are made.*

One area that most people forget about is the crankcase of an engine, this needs cooling just as much as the cylinder, it may not show up until you get a really hot day.

Two examples: My biplane design, "Yuppy Love" had a PAW 1.49 diesel, this engine overheated badly even though the cylinder was out in the open, I fitted a piece of tube to act as an airscoop to direct air over the crankcase, end of problem.

A friend has an ancient Junior 60 powered by an old O.S. 30, it has been flying regularly for 14 years. Suddenly it started playing up. The day was roasting hot, the engine was well carboned up on the outside but it

*Fig 7.3 – Cowling Sketch Page*

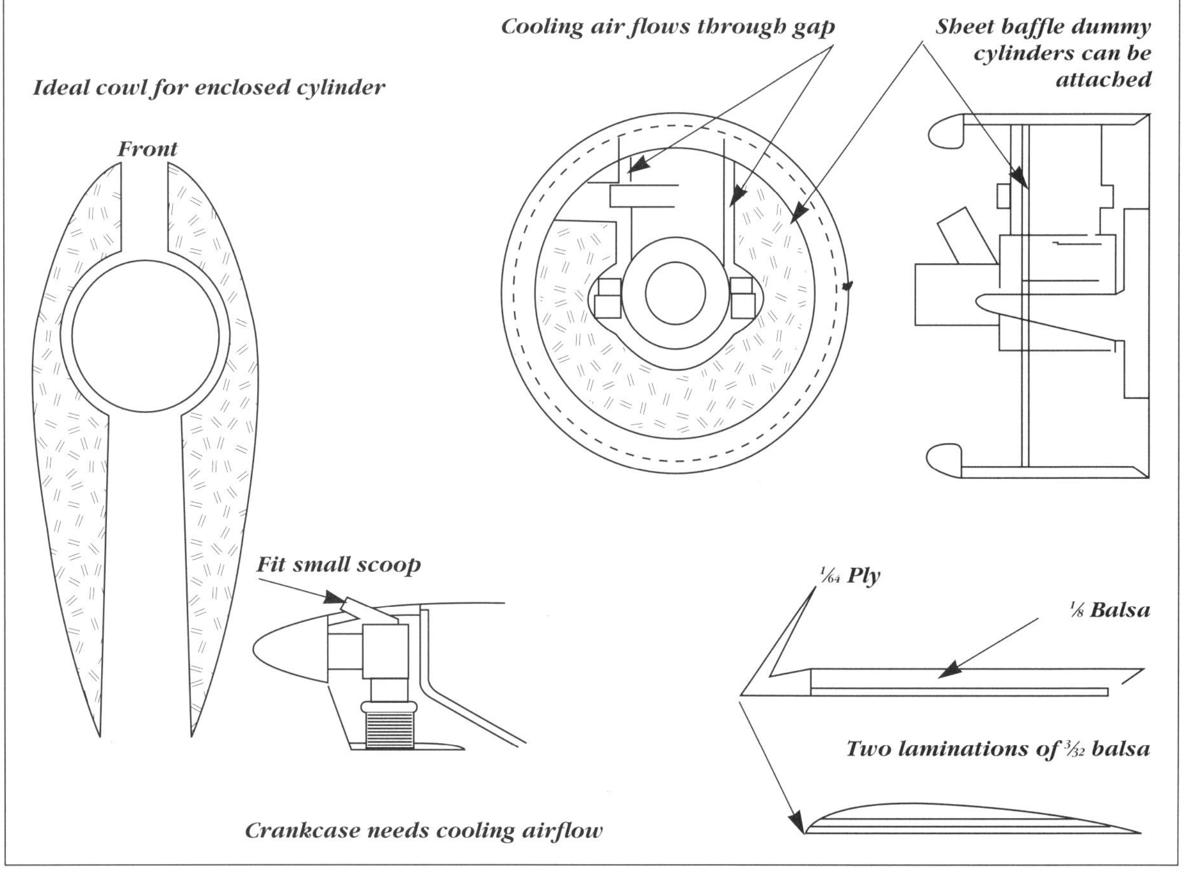

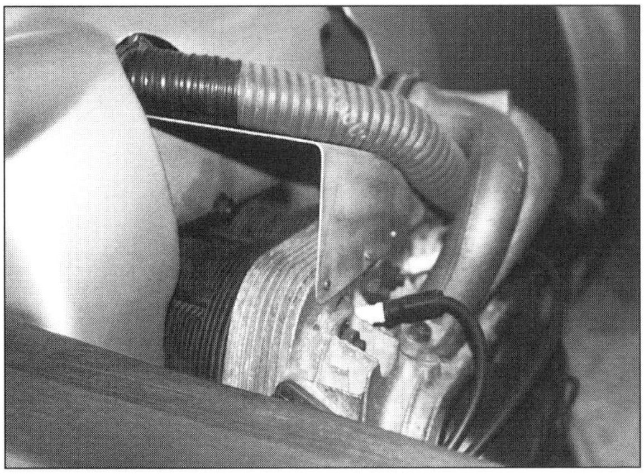

***This uncowled VW engine in a turbulent needs baffles to ensure that rear cylinder is properly cooled.***

had been flown in this state and these conditions before. What was different? My friend had fitted a spinner which covered the slot between the crankshaft and the cowl, cutting off the flow of air to the crankcase. Instant overheating.

The reason is that the incoming charge is pre-heated in the crankcase, this reduces efficiency but also sends hot mixture into the cylinder, that mixture is part of the cooling of the engine, the engine gets hotter, the uncooled crankcase gets even hotter and soon the vicious circle builds up to the point where the engine will not run.

One of the hardest things to design is a cowl that can be removed easily and yet will not come loose in flight. Hardwood blocks and self tapping screws can be one solution while cycle spokes can also be used to good advantage.

One thing to remember, cowls are going to hit the ground first and they will break regardless and they are tertiary structure anyway and so do not need to be tremendously strong.

Having said that, the radial cowl shown in the drawing is remarkably tough, I had one on my first Tequila Sunrise design, when that model crashed it smashed the carb back which broke the crankcase and bent the carb spigot. The cowl was undamaged.

Cowlings can also be moulded in plastic in the same way as canopies are.

## Canopies and Cockpits

Your sports model will look a lot better with a smart cockpit, this can be an open cockpit or a bubble canopy or some other sort of enclosure. This is a styling matter and will probably have little effect on the flight performance.

***High drag but good cooling on American Dragon.***

I did say 'Probably' but my Turbulent design (RCMW plans) has a big windscreen, if this got knocked off or badly broken I would have to wind in masses of trim on the elevator clevis to make the model fly properly and then take it out again when the windscreen was replaced.

In the case of a moulded canopy the important thing is to design it to suit what is available, as there is nothing more frustrating than trying to find one that fits what you have drawn, indeed you might even have to mould your own.

***The simplest form of cowl, firmly attached, good access and good cooling. This line is from Gerard Feeney "Callisto" design.***

The use of vacuum forming plastic allows you to design any shape of canopy, it also allows you to produce involved and tight fitting cowlings, spats and fairings.

Most people avoid this technique because of the need to make patterns. The other problem is finding somewhere that will do the forming for you but this is not as difficult as you might imagine.

The plug or pattern can be made out of hard balsa if it is only to be used for one or two parts, harder material would be in order for production runs.

***Monocoupe had very small outlet area so front of cowling is blanked off, model was still very fast on an O.S. FP .15.***

The finish is important, especially for canopies as these need to be crystal clear. Most normal finishes react to heat and so the wood grain shows through on the very first draw, this does not matter on a thick cowl but a canopy will be ruined.

I have found that M.D. Products Filler will give a perfect finish and the heat does not raise the grain of the wood, result, excellent finish for at least two formings and then another application of the filler will restore the finish. The pattern must be slightly smaller than the finished part as it is inside after the forming process. To get it out there must be no undercuts but rather a slight taper to aid removal of the pattern.

***A similar arrangement with Townend ring on Trivial Pursuit. No problem with outlet here.***

Radial cowlings can be removed by drilling a hole in the centre and applying some air pressure to this which will blow the pattern out of the cowl.

There are home made vacuum formers which use the suction from a vacuum cleaner, quite frankly these are not much use for a large, thick cowl or canopy; (A) the suction is not strong enough and (B) it is difficult to get the whole of a large sheet of plastic up to the right temperature. So where do we go from here?

Back to school. Most secondary schools or upper schools and even some middle schools will have a very good vacuum forming machine in their Craft Design and Technology Department. This will be capable of forming thick plastic about 10" x 18" which is big enough for most normal model uses.

You might wonder how to get the use of one of these machines. Of course if you have children there is no problem at all, if you don't, well, there are ways.

***Tequila Sunrise cowl survived when an engine was written off inside it.***

*Isaacs Fury Lycoming engine has baffles like the VW engine.*

*Tiger Moth air inlet is not very large. Internal baffling directs air through fins and between cylinders.*

All CDT departments are short of money, they are always grateful for materials which may be scrap or off-cuts, timber, metal, plastics, good cardboard, electronic components, anything to help the pupils make their projects. If you can supply anything in those lines you will be welcome. Even the offer of good old fashioned money.

To approach the head of the department you should telephone just after the lunch or final bell has gone when he is still there but has got rid of the pests, I mean pupils. You will NOT be popular if you phone during classes, the school receptionist will tell you the best time to phone.

*Tiger Moth air outlets are large, and there is one on each side and another underneath. To allow for the heated and expanded volume of air.*

***Open cockpits can look really smart on the right model.***

You may have to wait for your moulding to be done, the department may not have a full time technician who is the person who will be lumbered with the job. Be patient, remember they are doing you a favour and teachers are very busy people. I know, I used to be a school technician. You may also need to supply your own materials although schools normally stock polystyrene in 1mm and 2mm thicknesses. They rarely stock clear plastic for moulding.

***The Auster's Gipsy engine is hung on these bearers. I have seen heavier bearers on models!***

Failing the school, you could try the Yellow Pages, but these companies may not have a small enough machine and may not want to do a small one-off job. Some of the model manufacturers such as Chart may be willing to help out.

All this may seem a lot of hassle but a well moulded cowl or canopy can make a model stand out from all the others.

***Canopies have to be moulded which can present problems.***

# Chapter 8

# *Scaling Up Drawings*

There are two reasons for scaling up drawings. One is to enlarge 3 view drawings to a suitable size and then design the structure for a flying model, the second reason which is not really classed as designing is to enlarge a plan to suit a different size of engine, this is most common with vintage models and it will be discussed here because it is not a straightforward 'Blow it up two times and let's build it' process.

## *Scaling Up for Larger Engines*

You may come across a plan that you just have to build, for example a full size plan in a magazine that is designed for small engines and micro R/C, and which you would like to build for your .40 engine.

Generally speaking this is a pretty simple job with modern photocopiers because the enlargement factor is only going to be about x2 at the most. Many of my own small models have been enlarged by modellers in this way and all of them have been highly delighted. One modeller even took my Trivial Pursuit up to 10 foot span and built it in cardboard, you can now buy a kit of the development of this model . . . I can't comment on the kit because they never had the courtesy to let me have one.

Most full zoom photocopiers will enlarge over a wide range and definitely up to x2. All that you need to do is decide on the enlargement factor.

Now, assuming that you want to go twice size you just cut your plan up into A5 size pieces and get it taken up to A3, Bingo, you are there. Most pages are A4 and if you take A4 up to A3 that will be 1.41 times, which is pretty close to 1½ times enlargement.

Another way to look at it is this, decide on the span desired for the larger model, let's say 55". Now look at the actual span of the original, let's say 33". Divide 55 by 33 = 1.66 times. A good photocopier will cope with that.

Now it is not just a matter of sticking the bits of the plan together and building it because your material sizes are going to be ridiculous, take a 3/16" sq. spar enlarged twice, it will be 3/8" sq when a 1/4" sq. spar is the correct size.

***Trivial Pursuit has been scaled up for .25 sized engines and even enlarged to 10 ft wingspan and kitted with modified construction and slightly altered tail surfaces.***

***American Dragon is a 30" span for .8cc engines but has been scaled up to 60" span and fitted with aileron. Modeller was delighted with performance and handling.***

***New Zealand modellers scaled Tequila Sunrise up .40 and .60 engines, .40 version was modeller's first low wing plane.***

The best way to deal with spar sizes is to consider the cross-section, a 3/16" sq. spar is 3 x 3 = 9 sq 1/16", a 1/4" sq. spar is 4 x 4 = 16 sq 1/16" and a 3/8" sq spar is 6 x 6 = 36 sq. 1/16ths, four times the cross-section of the original spar but the 1/4" sq. spar is just under twice the cross-sectional area.

Sheet sizes are the same, 1/16" sheet on the L.E. can remain the same size, the sides should be increased to the size normally used for that size of model, so 1/16" sheet sides would go up to 1/8" but 3/32" side would still only go up to 1/8th.

The result of all this is a model that should weigh about two to three times the original weight for four times the wing area, but if you double all material sizes it will weigh 8 times the original's weight for only four times the wing area. Which will fly better? Which will be stronger? Go back and read chapter 4 again.

As a practical example, I took my 0-Fournier which is a 56" span powered glider with a Cox Texaco .049 and with three channel micro R/C and which weighed about 25 ounces, I scaled it up on a photocopier 1.41 times, I reduced the resulting 79" span to 73 to make the best use of standard wood sizes and finished up with a model that weighed 52 ounces, twice the weight, twice the area. I fitted a Super Custom 12 for power and four channel radio using small but not micro servos.

## Enlarging 3-View Drawings

The average 3 view drawing is either 1/72nd or 1/48th, there are other scales and the same principles apply. Most R/C scale models are from 1/7th scale upwards, this means that to enlarge a 3 view to model size we are looking at factors of 7 times upwards.

There are companies who can do enlargements to these factors but let me tell you, it is a very expensive process. You could do multiple photocopies but again, it would be expensive and all photocopiers distort so at the end you would have some rather non scale shapes. You would also have some very thick lines.

Believe it or not, the most accurate way is to scale the drawing up by hand. You can enlarge the 3 view slightly so that you have an exact enlargement factor because it is easier to enlarge 5 times than it is to enlarge 5.76 times.

***Fig 8.1 – The Effect of Doubling Material Sizes***

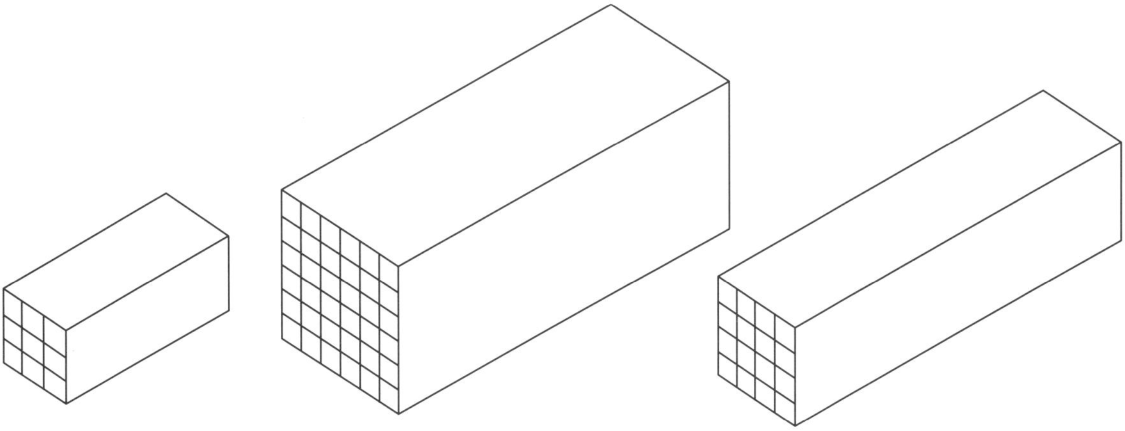

***3/16 Sq.***

***3/8 Sq.***
***All dimensions doubled***
***Cross-section area = 4 times***
***Volume = 8 times***

***1/4 Sq.***
***Length double***
***Cross-section x 1.3***
***Cross-section area = 1.77 times***
***volume = 3.55 times***

***O-Fournier powered glider has been enlarged 1.41 times with twice the wing area and exactly twice the weight.***

***Proportional dividers are expensive but save hours of tedious work, this pair is 30 years old.***

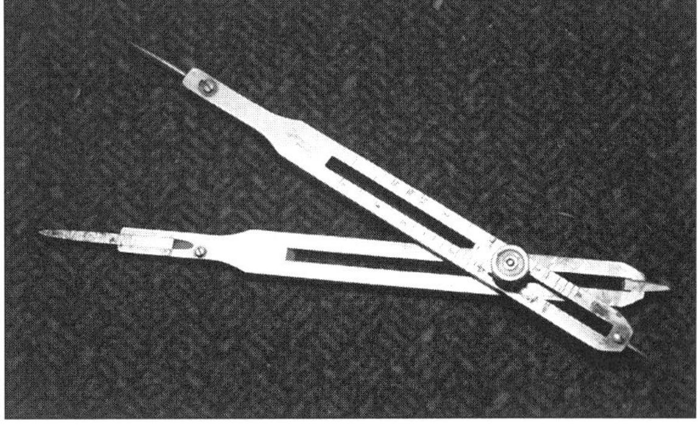

*Fig 8.2 – Scaling up a 3 View*

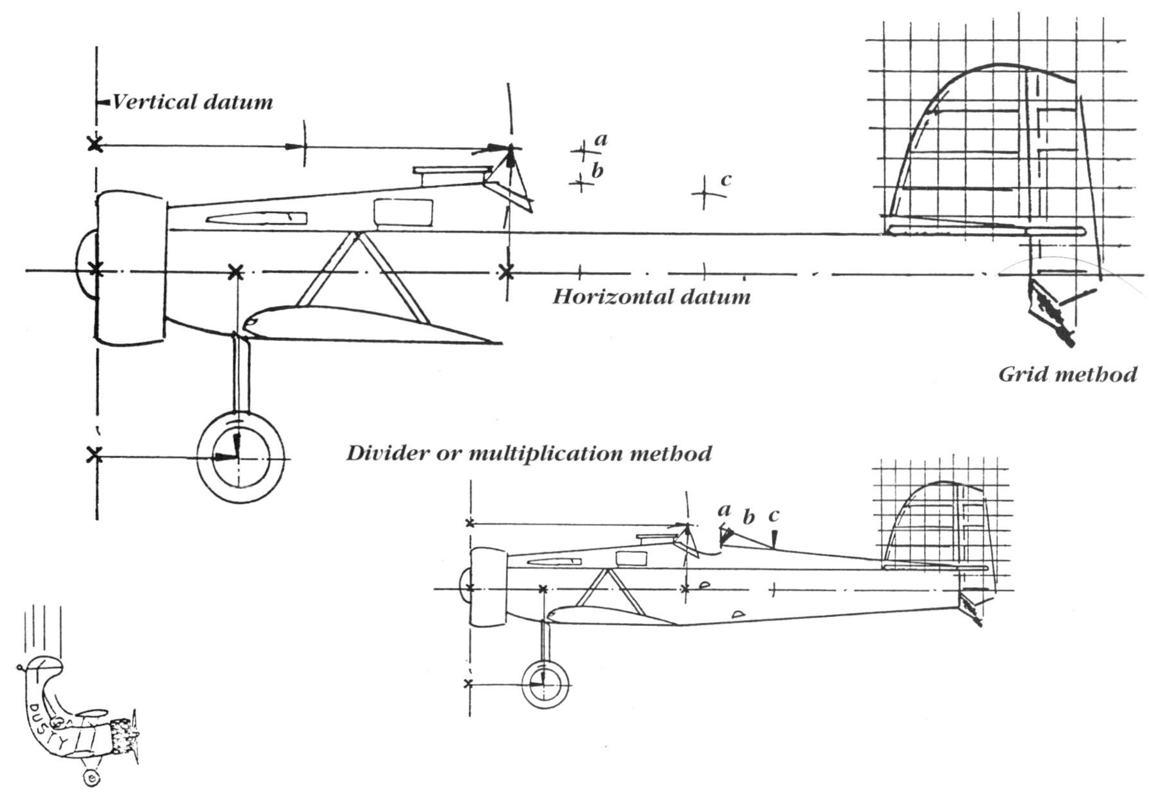


You decide on the span of the finished model, divide it by the span of the drawing which gives you 5.76. No good. Divide the model span by 5 and divide the result by the span of the 3 view and that gives the enlargement factor for the 3 view.

## Tools for the Task

Enlarging with a ruler and calculator is not on, there is an easier way. You use dividers.

Set your dividers on one dimension, say the spinner ring to L.E. on the 3 view. Now using your dividers, step that measurement off on the datum line of your drawing.

Now take the measurement from the datum to the chord line at the L.E. and step that off on your drawing. All main points are plotted in this way and then connected by lines until you have completed the outline of all parts of the aircraft.

I can hear screams of, "It will take a year to do that." No it doesn't, you will be surprised how much you can get done in a couple of hours once you get the hang of it.

You can save a lot of time and increase accuracy if you buy a pair of Proportional Dividers, these have a sliding joint and a scale and can be set to any enlargement factor from 1 to 10. They are very, very expensive, up to £30 last time I looked, but they are a tool that saves hours of work and will last for years, mine are still going strong after some 30 years.

You can buy cheaper ones made from two pieces of wire . . . you get exactly what you pay for in terms of quality, accuracy and ease of use.

## Enlarging

We have talked about the basic method of enlarging with dividers and this is pretty standard for the complete model but there are a few places where different techniques are needed.

Plotting the shape of a elliptical wing tip or fin and rudder can be done by plotting several points and joining them up with a French Curve, but you can also draw grids over the shape and a similar grid on your own plan and use this to plot the shape. Many American scale drawings actually had these grids printed on the plan, Nye and Wylam drawings for example.

***Possibly the ultimate design challenge, draw your own 3 views from full size plans and photos and then design the model. This is the author's Mong Sport control line model.***

***Taylor Monoplane, the author prepared scale 3 views from full size construction drawings and then designed model. With scale homebuilts you can follow the original construction which is often just like a model.***

Most plans provide fuselage cross-sections, in fact it isn't much use if these are not shown. Unfortunately these sections are rarely just where we want them and even worse there are not enough of them, so we have to plot the intermediate sections, you are going to have fun now.

The method I use works well. Plot and draw the largest cross-section and then the smallest on top of it. You will have four common lines connecting these sections, the datum lines.

Using a protractor plot more lines from the centre outwards. Now divide the lines into the required number of sections between the two existing ones and connect these points and you will have the intermediate sections which will be your formers. Cut them a little oversize just to be on the safe side. When you have done this for a couple of models you will know why I build flat sided models.

You can have some of the smaller parts enlarged on a photocopier but do remember that the lines will be very thick and there may be distortion.

One method which I have not tried is to take a 35mm slide of the 3 view and project it onto a sheet of paper. You will need a superb camera to get the quality and you will still have very thick lines.

I would suggest that you choose a nice simple subject with lots of straight lines for your first attempt at scaling up a 3 view drawing, there will still be plenty to do but the shape will appear quite quickly which is rewarding and there is less chance of errors and even these will be seen more easily.

## Scale Construction

There is no difference between a scale model and a sports model as far as construction goes, the scale model may have a more complex shape which will call for some of the fancier techniques but it is still just a model aircraft.

The average scale aircraft has a larger fuselage in relation to the wing area than the equivalent size sports model, this can add weight which is bad news. There is a natural tendency to want to add detail to a scale model, this will also add weight and is even more bad news. Finally, the colour scheme on a scale model may tempt you to do a fancy paint scheme and this will add yet more weight which is even worse news.

Bear all this in mind and make a conscious effort to save weight and you might keep the model somewhere within reasonable limits but it is difficult. Console yourself with the thought that if there were no weight limits at all for competition scale models, the builders would still be moaning that they couldn't keep within the limit.

***Fig 8.3 – Plotting Formers from Two Known Sections***

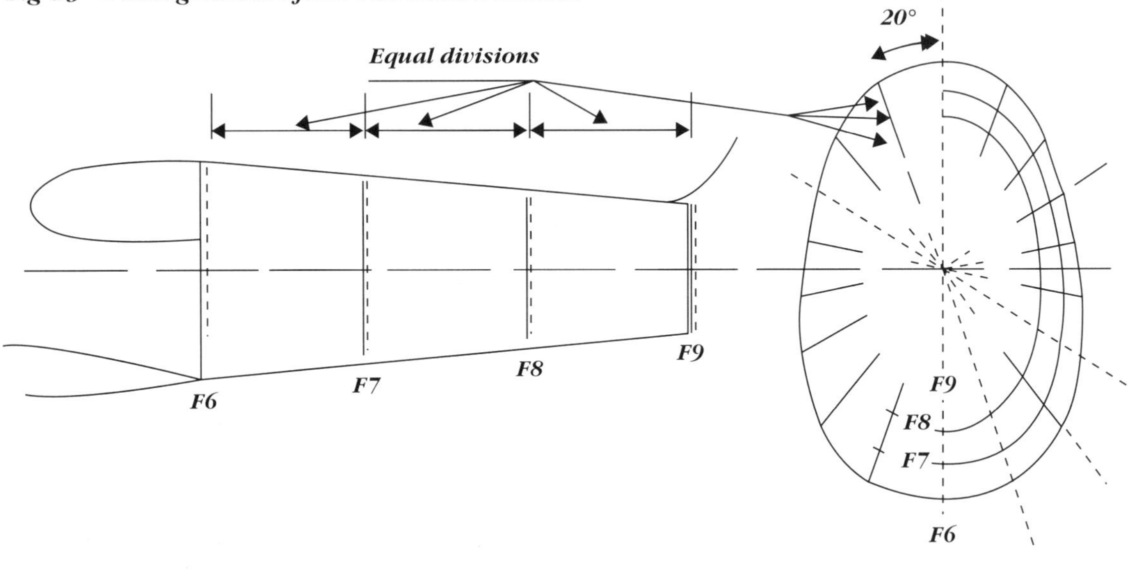


***F6 & F9 known F7 & F8 plotted***

***Plot points on radii and connect with French curve***

# Chapter 9

# Flight Testing

You stand in the workshop admiring your first ever own design model, you picture it taking off and flying round, you imagine the aerobatics and then the approach and greased on landing.

This probably takes place at about ten o'clock on a Saturday night and you are going to fly her in the morning. You are? Well you have quite a bit of work to do before that.

## Final Checkout

The first thing to check on is the CG, don't take "about right" as good enough, make sure that it is where you wanted it and preferably a bit further forward.

With a small tailplane the CG wants to be well forward, I favour starting with a CG at 25% chord for such models and going back as far as 30% for aerobatic models with large tailplanes, any further back is asking for trouble on a first flight.

Check that the incidences are as you designed them you should have checked this during building but check again. I always remember the old RAF safety poster which said "Don't Assume, CHECK".

You will have set up the control throws, check them again and check for correct sense. I like to set full rates at a little more than I think the model needs and low rates at a little less, I take-off on low rates. If you have exponential rates these are even better.

Check the undercarriage for correct tracking and if you have nosewheel steering, check that, I know that it is OK, I had a model and I knew that it was OK, fortunately I checked because the steering was reversed on the nose wheel.

Now charge the radio and hope for decent weather.

***Don Kichot, ¼ scale Polish homebuilt cruises past, note up trim needed for slow flight. A lovely model to fly.***

## Flight Test Stage One

One is always a little nervous before a first flight and this really is a first flight. You have a completely new model, not only has it never flown but no one has ever flown this particular design.

The worst thing you can do is to worry about things. Remember, it is only a model and if it is a conventional shape and it follows the rules set out in this series and has been built according to plan there is no reason why it will not fly.

Having said that you must give the model every chance because it can be badly out of trim, the CG could be a fraction too far back and the engine might decide to play up at the wrong moment so be careful.

Don't attempt to fly if the wind is too strong for your liking, I have a wind meter and I will not test fly in winds of over 12 mph even if the model is a proven design, it is better to wait for two or three weeks than write a new model off.

Do a range check on the field and get someone to double check that the controls are working in the correct direction. Run the engine up and hold the nose up, be very, very sure that the engine is going to keep going and that it will respond to throttle properly.

Some less experienced modellers get another pilot to fly their models for the first time. That is up to you but be sure that the pilot that you choose is experienced. Experience is not just a matter of being a good pilot, for test flying it also means being able to recognise a problem instantly and take the correct action, for example, if the model goes into a spin and will not recover does your choice of pilot know what to try in an attempt to save the model?

***Taylor Monoplane needed a dab of full down for spin recovery. It takes expensively earned experience to know what to do at fifty feet.***

***Fantasy (never sent for publication) flew like a dream. First model that I ever flew inverted at 2 foot altitude several times.***

## *The First Flight*

It is often a good idea to fly another model first, just to get relaxed and feel the conditions, you will have brought a second model just in case anyway.

It is a good idea to ask the other modellers for a slot to yourself for the first flight or at least ask them to fly well clear of the take-off area. If someone has a video camera ask them to film the flight for you, this can be useful later.

Now, go through all your fuelling and start up procedures calmly and carefully, have the nose of the model raised to check again for correct running, throttle back and have the model placed on the strip. Check that the controls move.

When you are satisfied that all is well, look around, open the throttle and take-off. Don't try and haul the model off the ground too soon, hold it down and get as much speed as possible before easing in up elevator. Be gentle, at this stage you don't know how the model will react. Climb to a good height making any turns slowly.

You will already have some idea as to how sensitive the model is to the controls, if it is very sensitive to elevator it would pay to throttle back and land and either move the CG forward or reduce the elevator throw or both. If the ailerons are too sensitive you can land and adjust the throws or carry on testing, sensitive ailerons are not as dangerous as elevators which can cause stalls, spins or flicks.

If the controls are sluggish you can go to high rates and see how the model performs in that state. Control sensitivity is a matter of personal taste. I like very powerful ailerons which other club members have trouble coping with.

Once you are happy that the model is in trim and flies reasonably you should gain some height, yes, more height, and throttle back, slowly feed in up elevator and watch for any tendency to drop a wing. Feel out the low speed handling because any vicious habits will catch you out on landing if you are not aware of them.

While you still have plenty of height (regain it if necessary), and are flying slowly with lots of up elevator, hit full rudder, the model should drop into a spin, allow it to turn three times and then centralise the controls, the model should stop spinning, if it does not, give a dab of full down elevator which should stop the spin, opposite rudder may also be needed. You can also open the throttle, if none of these stop the spin you are in deep trouble, but I have yet to find a model that did not stop spinning with down elevator.

Once the spin has stopped be careful and pull out as gently as the remaining height will allow because if the model was reluctant to recover from the spin there is a risk that it could go back into the spin as you pull out.

The cures for a model that does not want to stop spinning are, CG forward, add washout or, to be more drastic, a bigger fin and rudder.

If your nerves will stand it, you can now try loops and rolls and then landings.

***George Strohm's experimental model was fine at high speed but deadly as soon as it was throttled back.***

***First Tequila Sunrise flew well but had poor spin recovery due to poor choice of wing section, shape and wing loading.***

## *The Test Programme*

The elation of having made the first flight with your first design will have you walking around about six inches above the round and you will want to fly the model again and again.

The first thing to do is examine the model, just as you would any new model but with even more care as you will want to be certain that the structure that you designed is up to the stresses and strains imposed on it.

In subsequent flights you can fly through aerobatics but while you are enjoying yourself you should note if the model rolls with equal speed in both directions, if loops and bunts are equal, how much down trim is needed in inverted flight.

Now you can evaluate your model. Is it as aerobatic as you had hoped, is it unstable, do all the controls have roughly the same power? In short, are you happy with the model.

Get other people to fly it and see what they say. You may not agree with them but you are getting feedback which can be helpful for your next design. You may want to try some modifications, bigger control surfaces or perhaps cutting them down, maybe you will try fitting a bigger engine. The most likely thing that you will do is just carry on flying the model without any major changes but with the knowledge of things that you will change on your next design, because the best way to learn to design is by practical experience.

The more that you fly your new design the

***Analyse what went wrong***

***Second version of Tequila Sunrise was excellent and has been published here and in the USA and enlarged with success in New Zealand.***

***One scale up that didn't work, Aztec Two-step scaled up to 45" span for two .12s, came out grossly overweight at 60 ounces, tip stalled and flicked as soon as it was clear of the ground.***

more you will learn about it from the aerodynamic and structural aspects.

One word of warning, never get complacent. No matter how many models you design each model is different and you can still have failures.

When you do have a failure, analyse what went wrong because you can learn a great deal from your failures.

Failures can be divided into three main groups, aerodynamic, structural or pilot error, we will look at each of these in turn.

## Pilot Error

This is the most contentious group because if the model is marginal the pilot will usually blame the model and drawing the line between a situation that a pilot could have saved and one he could not can cause bad feeling between people. This is the reason that I always test fly my own models and hate being asked to test fly other people's models.

It must be said that a well designed model, properly set-up and flown by an experienced pilot should be safe, so one of the other failure groups will almost certainly be a contributing factor in any failure.

## Aerodynamic Failure

Most errors in aerodynamic design will show up immediately after take-off. Ask yourself exactly what your model did. Did it stall, did it flick roll or did it just refuse to lift off?

Stalling or flick rolling on take-off indicates a CG that is too far back or that the model had not reached flying speed before being hauled off the ground and this, in turn can be caused by too high a wing loading.

Aerodynamic failure can be caused by warps, incorrect angles, wrong CG, excessive wing loading, poor choice of wing section or combinations of these.

If this book has been followed and the model checked carefully then any errors should be of a minor nature and failures should not be a catastrophe.

## Structural Failure

These can happen and when they happen at height there is nothing that you can do except stand back and enjoy the crash.

Structural failures are rarely caused by major design faults, it is normally the smaller things like a wing attachment that has not been reinforced sufficiently or the use of incorrect grades of wood.

Failures after a heavy landing or minor crash will often show up weak points in the construction, very often caused by that sudden change in section.

The important thing is to learn from these things and go back to the drawing board and design a new model or modify the last one.

# Chapter 10

# Selling Your Plan

There are three monthly R/C magazines and about five bi-monthly magazines in Britain which publish at least one plan in each issue, most of them will publish two plans.

This adds up to a lot of plans every year. Some of the magazines will want only ducted fan or electric powered models, or perhaps gliders but they still want plans.

Where do they get these plans? From modellers like you, people who have designed a model and sent it in.

Do they buy every plan submitted? No, but if you send the right design, properly prepared and presented you have a very good chance that your plan will be published.

## What Sort of Design

Just about any type, but it should have something to attract the editor and potential builders. It can look beautiful, it can be ultra cheap to build or very quick to build.

Scale models are always popular but do call for careful selection of a subject, prototypes that have been done to death are less likely to be accepted but equally the very obscure type for which there is no documentation would need something special. Home builts are always popular.

The one important factor is that it must fly well, if the performance of your model is poor or it has very nasty habits, don't send it in because it will do nothing for your reputation and could affect the sale of future plans.

My first Tequila Sunrise was pretty good but it would flick into a spin at the slightest provocation and it needed a lot of height to get it out, pretty soon it came to grief because of that. I designed a new model which was fine and has been published here and in the USA.

Sometimes a design may have a characteristic that is dangerous but unavoidable. My Waco PG-2 (RCMW Plan) powered glider has two engines and no throttles, if one engine cuts you have to be very careful not to let it lose speed. In this case it is acceptable because it is a common fault with twins and you put that warning in the text. The Waco is a good subject, easy to build, good flying performance and can be flown as a glider or converted to electric power. Blue Movie was very quick to build, cheap, almost indestructible and flew superbly, the plan is a best seller.

## First Approach

You have a nice model which you think might be of interest to an editor. Choose one editor and send him a couple of photos and a description of your model and an SAE.

Do not send the same thing to more than one editor at a time because if they all want it . . . you have made yourself unpopular with most of them and that is not going to help you. If the editor is interested you have to be ready to send him the plans etc.

## The Plan

The plan needs to be neat and clear, it must show all the information regarding wood sizes and any instructions, all parts must be shown and identified.

***Blue Movie 2 (RCMW Plans Service), is a best seller because it is very cheap and quick to build and almost indestructible.***

***Ryan B-5 Brougham plans sold in the USA, aircraft is commercial version of Ryan NYP.***

***FILL THE FRAME. Note that all the static pictures in this book have the frame filled by the model, not like this one.***

***Flying shots will prove that your design flies, they do not have to be perfect.***

You must remember that many people will buy the plan without ever reading the article in the magazine so all the information needed to build the model must be on the plan.

Unless you are a skilled draughtsman your plan will have to be retraced to suit the standard and format used by the magazine. It will have to be traced onto Mylar film anyway for printing, the person who traces it cannot guess at information that is not included.

The plan that you send in can be done in pencil but I prefer to trace the plan and send in a print, this way you have not lost much if it gets lost in the post.

A roll of tracing paper is not too expensive and drawing pens can be very cheap for the disposable types, better ones are expensive but well worth it once you get into designing.

Some of my early plans were traced using the old springbow pens, which could be set to suit the line being drawn. Now I use 4 pens, .25mm for fine detail and grain marking, .5mm for most of the drawing, .7mm for main outlines and 1mm for the border of the plan. This is purely for my own satisfaction, as the drawing will be retraced anyway I suggest that you settle for .5mm which is an ideal size of pen.

***Tiny Shrike Commander is a hairy little beast but great fun and cheap, text warns of pitfalls.***

***This is better, my neighbour's drive gets used for most of my static shots.***

Lettering can be done free hand so long as it is legible, you can use a stencil but that is very slow.

Just remember, the better your plan is, the easier it is for the tracer and the less likely he is to make mistakes.

## *Text*

Your text must be clear and cover all vital points. It should start with a short introduction and cover the construction in a logical way. It should then cover the flying side of the model.

Get someone to read through your text to see if they can spot anything that needs changing. You know what you mean but will other people.

The text should be typed with double spacing on one side of A4 paper, it should be printed quite dark if possible so that a computer scanner can read it.

If you can't type it up you can often find a tame typist at work who will do it for a box of chocolates or there are secretarial services who will do it for a small fee.

## *Photos*

You will need a good set of photos, ordinary gloss colour prints 6" x 4" are quite acceptable, always send plenty to allow the editor to select the ones he wants.

The pictures must be properly exposed and in focus, this is not hard to achieve with the average compact camera.

What is important is to get a good background, it should be as free of any clutter as possible and should contrast with the subject. Construction photos should be taken against a dark background and the completed model taken against one that contrasts with the TONE of the model. The photo may be printed in black and white so you should think in terms of tone, not colour.

I use a neighbour's tarmac drive for most of my photos while construction photos are taken with the model on a dark coloured bedspread.

The other rule for taking pictures is: "Fill the frame", there is nothing more useless than a tiny figure holding something in a vast expanse of countryside or a minute model with a caption telling you to look at some fine detail.

***Velie Monocoupe in flight, for really good in flight photos you need a 300mm telephoto lens, a good pilot or cameraman, and luck.***

***How many things are wrong with this picture? Now you know what to avoid.***

Flying shots are helpful but very difficult to take, if you do not have a good, powerful telephoto lens, maybe a friend can help out. Editor's do like to know that a model does fly.

When I sent my first plan to the USA I sent an 8mm cine film of the model in flight (long before the days of video) . . . it worked.

## *Sending the Material In*

Pack everything up in a good, stout, card backed envelope. The print can be sent rolled but I usually just put them in the envelope. Send it by registered or recorded delivery and then sit back and wait.

Very often an editor will let you know that he has received it but editors are busy people so you may not hear for some time. If you have heard nothing after about a month you could telephone to make sure that it has arrived.

Once your plan has been accepted for publication forget about it.

You may be lucky and see it published in a matter of months but you could have to wait years. Do not keep ringing up and asking because it will not endear you to the editor and you can't afford to offend them, at least, not if you want to sell them more plans. You can remind them after a year has passed but do not overdo things, just send in more plans.

When your plan is published you will have the excitement of seeing your work in print and shortly after that you will get a nice at cheque in the post.

***This is not a good background as it is distracting.***

## *Where Next?*

With one plan awaiting publication the urge to design more models will grip you, and why not, there is a big market out there.

Once you have had two or three plans published you can consider the American market. If you look through American plans catalogues you will find several plans from British designers that have appeared in magazines over here.

The reason for this is that there is very little trade in magazines across the Atlantic and even less in the way of plans, so it is worthwhile for the Americans to publish these plans.

You can offer a plan to an American magazine that has been published here but you should talk to the editor of the British magazine first. R/C Model World buys the copyright but will normally give you permission to send the plan to the USA, other magazines only buy First British Serial Rights so you still own the copyright, but it is only polite to tell the editor anyway.

You must also tell the American editor if the plan has been published over here, some American magazines may only want a plan on an exclusive basis.

Any plan that you offer to an American magazine has to have something a little special about it, after all, they have a lot of designers over there. Slightly unusual scale subjects go down well, out of the rut sports models or even unorthodox, providing that you can prove that it flies well.

Get hold of some American magazines and study their style and presentation. Make your approach to the editor and see how you get on.

## *Looking Back*

This book has tried to show you the simple, uncomplicated way of designing models, a subject which can, and does, fill books.

You now know the basic mechanics of design and enough of the very simple aerodynamics and structures to design successful models.

There are books which will teach you more about aerodynamics if you feel the need, there are books on designing scale models as well as sports models. Add these to your collection as they will always be useful reference sources but the most important thing that you need now is experience and you can only get that by designing models.

*You need a good pilot and guts to get this close, even with a 300mm lens.*

*Moondancer trick shot. The model is in flight but how did I get it? Simple, take the moon at night with a 500mm mirror lens, rewind the film and reload on exactly the same mark and take your pictures as normal, with the moon in a clear area of sky.*

*This is my design for a scale Russian powered glider, the MAK 15. It is CO2 powered free flight and published by SAMS, would be a winner as an R/C design.*

*Filling the frame and pin sharp, flying pictures like this are hard to get. Top and bottom pictures are Gemini model kits.*

# Useful Addresses

**Aviation Bookshop**
656 Holloway Road
London
N19 3PD
Tel: (0171) 272 3630
*Aerofoil sheets, modelling books and magazines, full size aviation books and magazines. Also second hand books and magazines. Get their catalogue.*

**Transair Pilot Shop**
50a Cambridge Street
London
SW1V 4QQ
Or:
**Fairoaks Airport**
Surrey
Tel:(01276) 858533 for orders and catalogues.
*Square protractors, books, videos, aviation novelties and for the full size pilot, all navigation aids, equipment, simulators etc.*

**Balsacraft (International) Ltd**
Norwich Road Industrial Estate
Watton
Norfolk
IP25 6DR
*Top quality wood selected to your specification at the best prices, also large range of accessories.*
*Real value for money, one stop mail order, 3 first class stamps for catalogue.*

# Notes

# Notes